Garden Futures
Designing with Nature

Vitra Design Museum
Wüstenrot Foundation

Garden Futures

Designing with Nature

Vitra
Design
Museum

8   Forewords
14   Introduction by the Curators

22–55   # Paradise

24   Gardens are not just idyllic refuges, but have always expressed the values of the time and culture in which they were created. With works by Lucas Cranach, Hieronymus Bosch, Georg Gerster, Vita Sackville-West, Gabriel Guévrékian, Thomas Church, Luis Barragán, Mirei Shigemori, and others.

Carolina Maddè is an art historian, Project Manager at the Vitra Design Museum, and Curator of G10 Projektraum, Darmstadt.

56–111   # Garden Politics

58   Picture Essay   **Dreaming of a Global Garden**
Our gardens are filled with foreign and exotic plants. Luke Keogh explores their journeys through the object that transported many of them: the Wardian case.

Dr. Luke Keogh is a historian and curator. He is a lecturer in history at Deakin University, and Senior Fellow in Australian garden history at the National Museum of Australia, Canberra.

64   Essay   **From Garden City to Post-Industrial Landscape**
Oliver Sukrow charts how gardens have always reflected an imaginary future, even acting as utopian repositories for a yearned-for relationship to nature.

Dr. Oliver Sukrow is a Vienna-based academic who is currently completing research on architectures and landscapes of health.

76   Essay   **Acres of Opportunity:**
How we became Guerilla Gardeners
In the late 1960s, parts of built-up New York City fell into rapid decay. In an article that Liz Christy and Donald Loggins wrote in the 1980s, they explain how gardening became a tool of empowerment and a means to reclaim urban space.

Liz Christy was an artist and activist who is credited with founding the guerilla gardening movement. She passed away in 1985.

Donald Loggins is a photographer from Brooklyn who was part of the Green Guerillas when they formed in 1973.

82   Picture Essay   **Gardening in Times of Crises**
Jochen Eisenbrand documents how gardens have long played a central role in times of war, crisis, and displacement – as places of refuge, to ensure food security, and as symbols of patriotism.

Dr. Jochen Eisenbrand is a cultural scientist and Chief Curator at the Vitra Design Museum.

90   Essay

On the Lawn
Expanses of manicured lawn are an interface between humans and nature, but also a spatial expression of socio-economic status and control, recounts Kris Kozlowski Moore.

Kris Kozlowski Moore is a writer and editor based in London. His work has appeared in publications including *Real Review* and *American Suburb X*.

100   Picture Essay

Nature's Helpers
When engineers, graphic artists, or industrial designers set out to optimize the garden, they have one main goal: less work for the gardener, says Jochen Eisenbrand.

Jochen Eisenbrand

106   Picture Essay

The Invention of Leisure – From Park Bench to Lawn Lounger
Nina Steinmüller explores how the change in social ideals can also be seen in the design of garden furniture.

Dr. Nina Steinmüller is an art historian and Collections Curator at the Vitra Design Museum.

112–161   Testing Grounds

114   Case Study #1

Gardens for Everyone
With her tireless engagement both in the garden and in publishing, Dutch landscape designer, plantswoman, and publisher Mien Ruys (1904–99) set out to democratize garden design.

Leo den Dulk is a garden historian. In 2003, he became co-publisher and editor of *Onze Eigen Tuin* (Our Own Garden), the magazine founded by Mien Ruys. He published the first biography of Mien Ruys in 2017.

120   Case Study #2

The Collaborator
The landscape architect Roberto Burle Marx (1909–94) not only modernized Brazilian garden design, but made a substantial contribution to the protection of the rainforest with his research into native flora.

Isabela Ono is Director of the Instituto Burle Marx in Rio de Janeiro.

Nina Steinmüller

126   Case Study #3

"My garden's boundaries are the horizon."
In the face of his own death, British artist and filmmaker Derek Jarman (1942–94) created a flourishing work of garden art in a location most people thought of as inhospitable: rooted in the shingle on the southeastern coast of England – next to a nuclear power plant.

Nina Steinmüller

132   Case Study #4

Gardening to the Cycles and Rhythms of Nature
A garden can form the basis of an entire company, as it does at Weleda (founded in 1921). Here, business interests are subject to the finely calibrated cycles of nature.

Dr. Astrid Sprenger is head of medicinal plant cultivation at the cosmetics and natural remedies manufacturer Weleda.

138   Case Study #5   **The Disturbances of the Garden – In the garden, one performs the act of possessing**
For the past 30 years, Jamaica Kincaid's Vermont garden has been more than a place of work and pleasure. For Kincaid (b. 1949), it is also a springboard to interrogate aspects of colonial history, cultural appropriation, and displacement.

Jamaica Kincaid is an Antiguan American novelist, essayist, gardener, and gardening writer.

144   Case Study #6   **Can Art Be Nature?**
Piet Oudolf (b. 1944) is one of the most renowned garden designers working today. Celebrated for his elaborate artistic planning, his designs embrace the beauty of all the seasons.

Hanno Rauterberg is deputy head of the arts section of the German weekly *Die Zeit*. His award-winning features and commentaries follow events in the world of art and architecture.

150   Case Study #7   **"It's okay to borrow from popular culture."**
At first glance, contemporary digital culture and gardening may seem to have little in common. Inspired by the popular computer game *Age of Empires*, artist Zheng Guogu (b. 1970) combined aspects of Chinese tradition with pop culture to create a sprawling garden in his home town of Yangjiang.

Yujia Bian is a researcher specializing in landscapes, architecture, and art.

156   Case Study #8   **"All of our work is about connecting."**
In Kuala Lumpur, one of the world's most densely built-up megacities, the civic engagement and community spirit of a group of citizens led to the Kebun-Kebun Bangsar initiative which repurposed wasteland into a thriving garden – thus setting a pioneering example.

Ng Sek San is a Kuala Lumpur-based landscape architect and activist who focuses on local design and simple solutions.

Viviane Stappmanns has degrees in journalism and design. She is a curator at the Vitra Design Museum.

162–217   **The World as a Garden**

164   Essay   **The Planetary Garden**
In his essay, gardener Gilles Clément proposes a new understanding of gardening – one that views humans not as the rulers of plant life but as actors within a complex system of living beings.

Gilles Clément is a gardener, botanist, entomologist, biologist, and writer. In his writing and landscape architecture projects, Clément promotes an approach to landscape management that embraces the entire planet.

168   Essay   **Taming the Taxonomy**
Christoph Miler views existing botanical classification systems to be the result of design. He poses the question: if we reframe conventional – perhaps outdated – perspectives on the plant world, could this help us to reconsider our relationship with nature?

Christoph Miler explores migration, technology, and the environment through essays, interviews, and experimental poetry. Together with Isabel Seiffert, he forms the design studio Offshore with a focus on research and design education.

174   Interview   **"If we destroy the forest, we risk everything."**
In northern Ethiopia, agricultural deforestation has left behind arid plains. For Dr. Alemayehu Wassie Eshete, the well-being of the region now hinges on those caring for the remaining pockets of biodiversity.

Dr. Alemayehu Wassie Eshete is an Ethiopian forest conservationist who has worked with local church communities on preservation projects for over 20 years.

Viviane Stappmanns

180 Interview

"Human habitats need to accommodate other living beings."
Landscape architect Céline Baumann discusses different conceptual models for the future in her field of work.

186 Portrait

"Let's call it biospheric urbanism."
All landscapes are interconnected on a shared, global scale, says Bas Smets. In order to create environments that are livable while keeping our planet in balance, new microclimates can be introduced with a long-term view to less intervention.

190 Portrait

"We have to shift the way Indigenous knowledge is valued."
Based on her many years of global research, Julia Watson believes that some of the most promising technologies to help tackle future climatic challenges have existed for centuries. Her mission is to introduce this ancient wisdom into contemporary landscape architecture practice.

194

A Garden of Ideas
Maria Heinrich, Marten Kuijpers, Viviane Stappmanns, and Lisa Dabscheck glean a collage of ideas around how to collaborate with each other and with nature in our cities, buildings, schools, and living environments.

With works by Full Grown Studio, Alexandra Kehayoglou, Marian van Aubel, Stefano Boeri Architetti, Alexandra Daisy Ginsberg, Lacaton & Vassal, Ryue Nishizawa, Edible Estates, Fritz Haeg, Forest Gardens, Dan Pearson, Midori Shintani, Catherine Mosbach, Cercle d'Art des Travailleurs de Plantation Congolaise, Toulou Keur, James Hitchmough, and others.

219 Select Bibliography
220 Index
224 Image Credits
226 Acknowledgements
228 Colophon

Céline Baumann is a Basel-based landscape architect who works predominantly on urban projects but also maintains a prolific practice as an artist and educator.

Viviane Stappmanns

Bas Smets is a landscape architect and engineer whose projects range from private gardens to territorial visions and infrastructural landscapes.

Dr. Lisa Dabscheck is an ethnographer, writer, and editor based in Munich. Her work explores culture, society, and identity.

Julia Watson is a landscape architect based in New York City. With her studio Lo—TEK, she has conducted a multi-year research project into sustainable, climate-resilient, nature-based technologies.

Lisa Dabscheck

Marten Kuijpers is senior researcher at the Nieuwe Instituut, Rotterdam. He is currently coordinating a three-year program to develop an archive for garden and landscape design in the Netherlands.

Maria Heinrich is an architectural researcher and designer working for the Nieuwe Instituut in Rotterdam. Her research focuses on material cycles and ecologies of care.

Viviane Stappmanns

Lisa Dabscheck

Cover image:
© Vitra Design Museum, Illustration: Lorenz Klingebiel and Dominik Krauss; based on the photo: Les Jardins de Marqueyssac, Dordogne, France, © Laugery

Forewords

Vitra Design Museum

Gardens are places of productivity, pleasure, and regeneration. They express identities, visions, and dreams. Their cultural history ranges from the medieval *hortus conclusus* to the courtyard gardens of Islamic architecture. In recent years, as we have become more aware of the risks of climate change and depleting biodiversity, gardens have seen an impressive renaissance. No longer merely a bourgeois idyll or romantic retreat, they have become places of experimentation for a more sustainable future, which is reflected in concepts for urban farms or vertical gardens, as well as in the related discourses on power structures, biodiversity, and other topics that characterize a progressive perspective on gardening. The revival of the garden is evident on our very own doorstep, as the Vitra Campus has become home to a garden created by one of the most influential exponents of the new garden movement, Dutch garden designer Piet Oudolf.

The garden has become an avant-garde space.

This sets the scene of *Garden Futures*, the first major exhibition on the history of modern garden design, encompassing cultural pasts, concepts of classical modernism, and contemporary projects that are delivering new ideas for sustainability, nutrition, health, and biodiversity. Offering an overview of the exhibition's wide range of exhibits and its themes through in-depth essays and texts, this book gives insights into contemporary gardening discourse, touching upon topics such as biodiversity, climate change, and the legacies of colonialism.

Our co-producer, the Nieuwe Instituut in Rotterdam, has played a crucial role in the curatorial development of the exhibition, providing a special focus on non-Western perspectives. The Wüstenrot Foundation, our other co-producing partner, contributed to an emphasis on the reform movements of the early twentieth century and to the idea of gardening as an integral part of a liveable urban future. The exhibition's Global Sponsor, Weleda, is a major manufacturer of natural cosmetics and medicinal products; it has been an enriching experience to discuss the cultivation of plants and herbs used for cosmetic and healing purposes with experts at the company, whose gardens are not only their main resource but a symbol for the company's identity that is closely connected to a holistic understanding of nature's cycles.

Our sincere thanks go to our partners and to all with whom we have had inspiring conversations, which broadened our understanding of the garden, opened new pathways of research, and added to the manifold perspectives included in both this book and the exhibition. We would also like to thank the curators Viviane Stappmanns, Marten Kuijpers, and Nina Steinmüller, who compiled an impressive range of material and translated a wealth of knowledge into a fascinating exhibition. Its curatorial narrative extends from the medieval metaphor of the garden as an earthly paradise – famously depicted in Hieronymus Bosch's *The Garden of Earthly Delights* – to a post-industrial perception of the whole planet as a garden, which we all must tend to. We sincerely hope that this exhibition will contribute to this endeavour.

Dr. Mateo Kries
Director of the Vitra Design Museum

Wüstenrot Foundation

Creators of gardens believe in the future. Gardens are also a very specific expression of what it means to be human at a certain time and in a certain culture. Are there any cultures without gardens?

The importance of gardens goes far beyond the romantic notion of a magnificent floral spectacle promising escape, self-discovery, or a balm for the soul. Gardens are rather a construct and a cultural artefact, the product of an idea, a dream, a vision – but always forever growing and decaying, that is, in a constant state of flux. It is something that gardens share with nature. Yet, they are also the very opposite of nature.

The exhibition *Garden Futures* has a lot of new things to say, in that it reflects on the meaning of the modern garden in the broader context of design, architecture, and urbanism – before now, a little heeded discourse. Countless examples are cited to illustrate how gardens have changed over time and to discuss how garden design is affected by political and cultural values. The exhibition also examines the role of the garden as a retreat, and its influence on architecture and urban planning. Last but by no means least, viewers are called on to consider planet Earth as one gigantic garden, and that it is up to us to take care of and tend to this planetary garden.

Gardens are increasing twofold as laboratories in which to find answers to the most pressing problems now facing us. Soaring population growth, inexorable urbanization, changing economic and social conditions, and spiralling individual needs are among our many challenges. The Wüstenrot Foundation is grateful to be part of this very special project that sets out to discover how the garden has become a catalyst for a better future. Key questions such as urbanization and sustainability preoccupy our daily work. Over a wide range of projects, we are exploring what the solutions to the challenges of tomorrow might look like. We regard cultural heritage, of which the garden is obviously a part, as both a starting point and our compass in this endeavour.

The experience of the Covid-19 pandemic has given many of us a new and deeper appreciation of our own gardens. Yet, the risk that climate change and environmental pollution will destroy our natural surroundings is growing all the time. This exhibition presents impressive evidence of the political, historical, and cultural significance of gardens, without undermining the possibility of still cherishing them as secluded and romantic retreats and as havens of peace and privacy. May both the exhibition *Garden Futures* and this book help foster an appreciation of gardens as places of manifold benefits that are worthy of our protection and which deserve to be firmly anchored in our memory.

Prof. Philip Kurz
Managing Director of the Wüstenrot Foundation

Nieuwe Instituut

Very few of you likely know the Ecokathedraal, a wild garden in the tiny village of Mildam in the north of the Netherlands. In this garden, made by the unruly artist and self-proclaimed "ecotect" Louis le Roy (1924–2012), ecologists recently observed an incredibly rich ecosystem, seemingly far older than its actual age of 50 years. No one knows exactly how and why. Except for Le Roy. He knew what he was doing, mixing philosophical and ecological theory alike, arguing against the monotony of controlled, "rationalist" modes of land development that still hold sway today. His gardens are not so much designed; instead, they gradually self-propagated on a foundation of stacked construction waste. Residents are actively involved in the laying and sometimes the sowing and planting of these wild ruins. After that, maintenance is limited to what is absolutely necessary. Time and natural growth processes are given plenty of room.

Gardens are the ultimate example of where humanity's desire to control the environment clashes with a fascination with the wild, untouched forces of nature. Throughout history, gardens have reflected differing notions of beauty and expressed competing visions for better futures. The Netherlands has a rich tradition of garden and landscape design. Though better known for creating land from the sea and making it productive, the country's relationship to its landscape presents a far more nuanced picture when seen through the lens of Le Roy and others. Unfortunately, most archives of garden and landscape design in the Netherlands are inaccessible or even threatened. Together with various partners, the Nieuwe Instituut is investigating a long-term perspective for preserving and learning from these important legacies.

Along with reflecting on the past, we are a testing ground for future practices. Recently, the Nieuwe Instituut became a Zoöp, meaning we formally adopted an organizational model for the cooperation between human and non-human life to safeguard the interests of all *zoë* (Greek for "life") and involve them in our decision-making processes. The Zoöp model is indebted to the many Indigenous cultures that attribute personhood to non-human entities, and the New Garden, which since 2015 has been part of the outdoor space at the Nieuwe Instituut, is one key aspect, with biodiversity and ecological management at its core.

So when the Vitra Design Museum invited us to co-curate the exhibition *Garden Futures*, we were eager to accept. It offered the opportunity to not only show a great variety of international garden and landscape design, but to also reflect on these designs as testing grounds for more socially and ecologically just futures. Among other things, the show highlights historical as well as more recent and contemporary projects that go beyond human-centric approaches to offer a fuller range of possibilities for designing our relationships with nature. We hope it will inspire visitors to see beyond conventional assumptions about garden design, and explore the many ways in which designing *with* nature is actually possible as we work towards not just sustaining but healing our cities, landscapes, and the planet.

The Nieuwe Instituut is the Netherlands' national museum and institute for architecture, design, and digital culture. Based in Rotterdam, the institute's exhibitions, public programmes, research, and wide-reaching national and international initiatives provide a testing ground for collaboration with leading designers, thinkers, and diverse audiences, critically addressing the urgent questions of our past, present, and future. In addition to housing the National Collection for Dutch Architecture and Urban Planning, the Nieuwe Instituut manages the 1933 Sonneveld House, a leading example of Dutch functionalist architecture, as part of its campus in Rotterdam's Museumpark.

Aric Chen
General and Artistic Director of the Nieuwe Instituut

Weleda

Gardens count among humanity's earliest design feats. A fence or a hedge, a flower bed or two is all that is needed to turn a patch of untamed nature into a cultural space born of a specific intention. The design of a garden tells us a lot about where the designer stands in relation to nature: does it represent the imposition of an idea? Or is the idea rather subordinate to the ongoing process by which the garden itself evolves, that never-ending cycle of impulse-response-impulse-response?

We are currently designing the world to death. In our pursuit of short-term goals, we are trashing our oceans, changing our climate, stripping our planet of its biodiversity, and draining our soils of all life. Our hope is that enough people are now rethinking and seeking to find harmonious solutions with nature. Design can become a crucial force for good, which turns waste into new resources, avoids emissions, and promotes diversity, and that being embedded in the cycles of nature only enhances life itself.

Welcome to the Weleda Gardens!

For us at Weleda, the garden is where the relationship between humans and nature is most vividly apparent. The dialogue between the two is manifested in the garden as organism, in all its diversity. Gardens are nature's answer to human intentionality. And because nature always responds in its own way, garden design can never be more than a suggestion – perhaps even just a question.

Hence, our proposal that design stops wanting to be the answer. Design that seeks to be truly sustainable should rather take on the character of a question, so that change can become part of the design.

The garden is also the perfect place not only to pose questions, but, on close inspection, to find other answers – answers to the challenges that we ourselves have created. How do we handle scarce resources? How do we design sustainable processes? How do we make the environment an integral factor in our thinking? And how do we respond to the unforeseen? Gardens teach us that design is predicated on an interest in life in all its myriad forms.

Furthermore, it is interesting to observe that plants tend to grow especially well if we pay less attention to them and more to their habitat, above all, to the soil. This, too, is something that gardens teach us: design calls for a focus on context, not the subject.

We are deeply grateful to the Vitra Design Museum for having created an exhibition that stimulates a new perception of design, a design that exists in dialogue with nature, and which may ultimately save it. That we at Weleda are able to test these necessary principles in our own gardens every single day is a great gift.

Christoph Möldner
Head Global Communications Weleda

Imagining Futures Otherwise
or
Why We Should All Be Gardeners

Introduction

by
Viviane Stappmanns
Marten Kuijpers
Maria Heinrich

That our relationship with nature must change has never been clearer than today, with climate and ecological crises threatening humanity and the planet. Yet in the face of complex challenges, where do we find real, impactful answers – let alone a completely new attitude? The garden is a good place to start looking. Gardens present the world in miniature. What constitutes nature for us and how we relate to it – be it as individuals or as a society – are questions that, again and again, across all ages and cultures, have always been negotiated in the garden.

Every garden, no matter how big or small, has always been a place where futures were made and conceived. For centuries, gardens have been spaces on which to project our hopes and aspirations. The anxious amateur gardener huddles over small pots in the basement in late winter, sowing tomato seeds in the hope of a bountiful summer. The absolute French monarchs, planning vast gardens arranged in strict geometry around a centre of power, outlined a future – theirs – in which all living beings, even the plants, would be under their control.

Gardens reflect identities, dreams, and visions, and this is what *Garden Futures* is about – an exhibition curated by the Vitra Design Museum and the Nieuwe Instituut, alongside which this book was created. The recent revival of horticulture has focused less on the garden as a romantic refuge than as a place where concepts of social justice, biodiversity, and sustainability can be tried and tested. Gardens have become places of the avant-garde.

With the exhibition, we wanted to find out where today's garden ideals come from. But we also wanted to explore how gardens can help us achieve a liveable future for everyone.

In exploring the garden's past, its symbolism, and its significance for the future, it was inspiring to come across the work of like-minded researchers. In their book *Earth Perfect? Nature, Utopia and the Garden*, researchers Naomi Jacobs and Annette Giesecke argue that it is important to analyse and reconstruct the history, symbolism, and vital potency of gardens. And they go on to wonder: "Can a new ethos grounded in gardening lead us to a more sustainable relationship between humanity and the natural world?"[1]

We found much value in Jacobs's and Giesecke's suggestion to look at the past in order to learn for the present. The multilayered stories that garden design can tell us often reveal remarkable connections. The English landscape gardens of the eighteenth and nineteenth centuries, for example, presented images of untouched nature, but in reality, every blade of grass was meticulously planned as part of an orchestrated whole. "Like nature, only better", novelist and critic Jim Lewis remarks – better for humans, that is, or, in the words of Lewis "with all the awkward bits smoothed out".[2] Today, the lawns that originated in eighteenth-century England have morphed into monotonous green expanses engulfing public parks and suburban gardens all around the world. Landscape

architect James Hitchmough calls them "green deserts".[3] On their bare surfaces, the beliefs that have defined the Western world for hundreds of years lie exposed in startling nakedness, including the conviction that Man must have dominion over nature, and that nature is merely a playground for his activities.

Since the Vitra Design Museum and the Nieuwe Instituut are both institutions with a focus on the remit of design in the broadest sense, we were also guided by the question whether (new) practices of gardening, of caring for and working with nature, might hold the promise of a new attitude and practice of design as well.

Even contemporary design often has little to do with sustainability, instead functioning as "a commercial tool executed at the direction of clients or employers". This is a point raised by curators and critics Paola Antonelli and Alice Rawsthorn, who in their 2022 book *Design Emergency* suggest that design's "limited role to which it was confined in the industrial age" needs to be broadened to "build a better future not only for human beings, but also for the other species with whom we share this planet".[4]

← Bagh-e Shahzadeh, Persian garden near Mahan, Iran, 2014

The concept of the garden
as a healer and saviour is
a recurrent trope taking on
a wide variety of shapes.

Our research led us from the dominant concept of the Western garden as we knew it towards collective, traditional, and Indigenous practices of caring for nature, which are not usually discussed under the rubric of "gardening".

We discovered that even the earliest and most original concepts of the garden yielded valuable clues, given that the expressive power of gardens as a human-nature interface is particularly evident in myths and religions, in literature and the fine arts. The Islamic concept of the garden brings paradise to Earth – but only for a chosen few. Because of their layout, these gardens are known as *chahar bagh*, or "four gardens". The spread of Islam carried them from Ancient Persia across the Middle East to Mughal India, North Africa, and Mediterranean Europe, and their four-part structure found its way into other art forms, too, like miniature paintings or Persian carpets.[5] In Christianity, the Garden of Paradise is a crucial concept as well, as is the Garden of Eden. And here, too, nothing is for free. Access to the garden is a reward; exile is a punishment. In either case, the garden is a space of yearning and transcendence.[6]

The concept of the garden as a healer and saviour is a recurrent trope taking on a wide variety of shapes and forms across the centuries. In the industrialized West of the nineteenth century, it resurfaces in the allotment gardens promoted as a counterweight to urban squalor. In Germany, they bear the name of an ardent advocate, the physician and educator Gottlieb Moritz Schreber. While the Schreber gardens guaranteed a basic supply of food, they were also considered a moral institution: the sunlight, fresh air, and exercise provided by working in the garden was expected to counteract temptations towards drink or violence. The garden became a symbol of a better, greener, and healthier future.

However, while the garden never lost its power as an idyllic retreat, it is also a projection space and battlefield in which political and commercial interests, societal value systems, and individual desires clash.

"The rich man owns a garden," landscape architect and designer Barbara Stauffacher Solomon remarks, "the poor man works in it."[7] While she is referring to European aristocrats and their expansive gardens, the principle applies on a global scale as well.

Western colonial powers owned plantations, enslaved people and indentured labourers worked on them. The colonies also served as study sites for botanical inquiry which often requisitioned Indigenous knowledge of nature and the properties of plants. The notes of a select few who travelled in the name of science often throw a chilling light upon this. On her journey in 1699 to Suriname, entomologist and botanist Maria Sibylla Merian recorded that a plant she was studying, the Peacock flower, was used by enslaved women to induce abortion – to prevent children being born into slavery.[8]

The notion of exclusion defines the garden – quite literally. The word "paradise" originates from an ancient

Persian word meaning "walled enclosure",[9] and even today, most dictionary definitions suggest that what actually constitutes a garden is its separation from its surroundings by some kind of enclosure. A convent garden was a *hortus conclusus* – a contained garden. But all gardens were designed to keep out undesirable elements:[10] the faithless, naturally, but also non-human life – plants and animals – which is admitted only if it serves a purpose for the humans tending to the garden, adding to whatever they seek from it – sustenance, leisure, pleasure, healing, and – usually – some kind of escape.

Only a controlled nature can offer people sustenance and refuge: this paradigm needs revisiting. "Do we continue to nourish dreams of escaping," philosopher Bruno Latour inquires, "or do we start seeking a territory that we and our children can inhabit?"[11]

So whom can we learn from? Many twentieth-century landscape architects and gardeners spent a lifetime experimenting with the garden as an interface between culture and nature.

Brazilian landscape architect Roberto Burle Marx described the garden as "the interaction of man and nature", a place "where the right balance between the small interior world and the immensity of the exterior world recreates harmony and achieves serenity".[12] He explored this relationship in both his drawings and – on a grand scale – in the gardens he created, often using native plants he had discovered in the Brazilian rainforest.

Dutch garden designer Mien Ruys was a true plantswoman – and foregrounded issues of privilege and hierarchy in her work. Her books and her many articles in newspapers or her own magazine *Onze Eigen Tuin* (Our Own Garden) provided amateur gardeners who could not afford to hire a garden designer with instructions on how to design and plant their gardens.

Some theoretical approaches of the past, too, have lost none of their relevance. In 1966, landscape architect Ian McHarg – whose book *Design with Nature* continues to inspire landscape architecture students to this day – was part of a group who announced that

← Unknown gardener in the Monte Verità commune, Ascona, Switzerland, c. 1920

↗ *Das Volk im Zukunftsstaat* (The People in the Future State), Illustration by Friedrich Eduard Bilz, 1904

an age of environmental crisis was upon us. The ideas and approaches they formulated in their "Declaration of Concern" are as pertinent for design practice today as they were then: "There is no 'single solution'," they observed, but only "groups of solutions carefully related one to another." And they added: "There is no one-shot cure, nor single-purpose panacea, but the need for collaborative solutions."[13]

Particularly rewarding, too, is the framework provided by a contemporary voice: French botanist and gardener Gilles Clément. It was he who coined the term "planetary garden". As the garden is always an enclosure, according to Clément, it now encloses the whole planet, which introduces a new collective responsibility, endorsed in this book and the exhibition (Clément's essay can be found on page 164).[14]

Studio Céline Baumann even go so far as to propose a "Parliament of Plants". The idea is simple: what if human life forms were not the only ones to have agency in decision-making processes – what if democracy involved non-human species as well? In a "green democracy", Baumann believes decisions must necessarily be based on the common good, mutual care, and support.[15]

Many approaches we considered for the exhibition involved experiments and artworks that address and dissolve the boundaries between human-made and natural, sometimes incorporating artificial intelligence as a non-human perspective. Daisy Ginsberg's *Pollinator Pathmaker*, for example, is a tool for gardeners wishing to create a pollinator-friendly garden – but it also shows us the world from the point of view of bees, bugs, and butterflies. London-based ecoLogicStudio, meanwhile, taught a machine-learning algorithm to behave like living slime mould in order to depict the future of "bio-digital autonomous" cities overgrown by an organism that forms a symbiosis with their human inhabitants.

So whom can
we learn from?

For many gardeners, activists, and social entrepreneurs, the future – and in some cases, the present – lies in interconnected human and natural environments organized in eco-social networks. Here, the garden is no longer defined as an enclosed space; on the contrary, the ethos and practice of gardening spreads its tentacles across entire cities. In New York, the Green City Force is a social enterprise that trains young adults from disadvantaged communities in gardening and urban farming, tackles food insecurity and urban climate challenges – and forms part of a network of numerous governmental and non-governmental agencies and businesses committed to rethinking the status quo towards a new, regenerative, and inclusive economy.

Other researchers, artists, and practitioners are claiming that we need not look far into the future, but into the past to find relevant approaches. Forest gardeners view the existing biodiversity and greenery as an asset and part of their oeuvre. These gardens exist in different varieties and scales and with varying amounts of human interaction, but their origins are ancient. One such example is the Milpa Cycle, a Mayan agricultural technique that has been practised for 8000 years in Mesoamerica.[16] Landscape architect Julia Watson, with her practice Lo—TEK, has researched Indigenous ecological knowledge and agricultural techniques such as agroforestry or the Milpa Cycle and suggests incorporating these rich traditions into landscape design, where they could help develop new methods for achieving food sovereignty or creating more sustainable living environments.

For some landscape architects, whose daily work is to design the interface between humans and nature, one thing is clear: we need to find new ways and symbioses. It might help to transform cities into protected zones for plants and animals that are exposed to great dangers in the monocultures of industrial agriculture: this idea was proposed by Malaysian landscape architect Ng Sek San. Meanwhile, Catherine Mosbach and Bas Smets and their design offices are busy analysing and regenerating sites of past environmental transgressions for a new future. The defining design question, as Bas Smets puts it, aptly and succinctly, always being: "What would nature do?"[17]

So there's a lot to learn in the garden and in landscape design about how we can rethink and actively shape our relationship with nature. In our research we have found places, ancient traditions, gardeners, artists, and philosophers who act as role models and provide blueprints. No wonder, since the garden has always been a place of learning – whether morally charged or quite literally as a place of education.

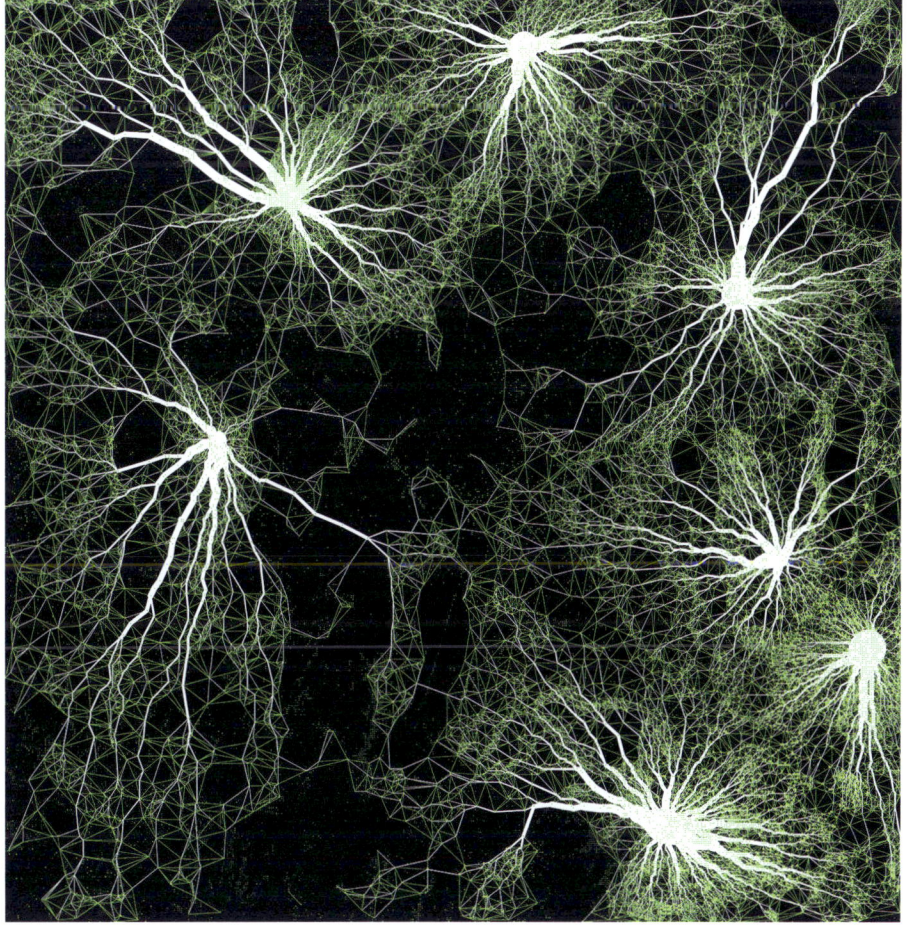

↖ Maria Sibylla Merian, Illustration of Flos Pavonis from *Metamorphosis insectorum Surinamensium, ofte verandering der Surinaamsche insecten* (The Metamorphosis of Surinamese Insects), Gerard Valk, Amsterdam 1705 (Coloured first edition)

← Cover of Ian McHarg, *Design with Nature*, 1971 edition (originally published 1969)

→ ecoLogicStudio, Deep-Green Biotic network analysis of path systems in Guatemala City, 2021

What would nature do?[17]

And yet, while *who, what, where* are salient questions, the most important question to ask is *how*?

In garden design, as opposed to other disciplines, collaboration has always taken centre stage. Gardening knowledge has long been shared from person to person, often informally across the garden fence. In our research, we have repeatedly come across networks of garden designers meeting regularly to share knowledge and to help and encourage one another. Likewise, gardening is always about working with local conditions, not taking too much, caring, and giving back.

Imagine a design practice that hinges on large collaborative networks: in the face of the complex ecological and social challenges we are facing, the idea is as compelling as it is promising.

More than that, garden design, unlike other design disciplines, is predicated on collaboration with non-humans. Take gardener and psychologist Sue Stuart-Smith, for example, who says: "I see gardening as a repetition: I do something, then nature does its part, then I respond, and so on, not unlike a conversation. There's no whispering or shouting or talking, but in that back-and-forth there's a delayed and sustained dialogue."[18] A new design practice could work in the same way, as a dialogue with nature and with each other, a balanced give-and-take, with room for failure. Maybe this kind of design could be less about finding grand solutions and more about starting small. Or, as music legend Brian Eno puts it: "Think like a gardener, not an architect: design beginnings, not endings."[19] ●

↑ Green City Force Eco Hub at Wagner Houses (part of the New York City Housing Authority), East Harlem, Manhattan, 2022

↗ Sue and Tom Stuart-Smith, Plant Library at the Orchard, Serge Hill Community Garden, Hertfordshire, 2022

1   Annette Giesecke and Naomi Jacobs, *Earth Perfect? Nature, Utopia and the Garden*. London: Black Dog Press, 2012, p. 9.

2   Cecily Brown and Jim Lewis, *The English Garden*. London: Karma, 2015, p. 9.

3   See page 217 of this volume.

4   Alice Rawsthorn and Paola Antonelli, *Design Emergency: Building a Better Future*. London: Phaidon, 2022, p. 11.

5   Hosna Pourhashemi, "Paradisiesches Abbild", *Tec21* journal (20 September 2018), pp. 25–33.

6   Annette Giesecke and Naomi Jacobs, *Earth Perfect?: Nature, Utopia and the Garden*. London: Black Dog Press, 2012, p. 7.

7   Barbara Stauffacher Solomon, *Green Architecture and the Agrarian Garden*. New York: Rizzoli, 1988, p. 57.

8   Maria Sibylla Merian, *Metamorphosis insectorum Surinamensium, ofte verandering der Surinaamsche insecten*. Amsterdam: Gerard Valk, 1705.

9   Nadine Olonetzky, *Inspirations: A Time Travel through Garden History*. Basel: Birkhäuser, 2017, p. 18.

10   Annette Giesecke and Naomi Jacobs, *Earth Perfect?: Nature, Utopia and the Garden*. London: Black Dog Press, 2012, p. 9.

11   Bruno Latour, *Down To Earth: Politics in the New Climatic Regime*. Medford: Polity Press, 2018. p. 5.

12   Marta Iris Monteiro, *Roberto Burle Marx: The Lyrical Landscape*. Berkeley, CA: University of California Press, 2001, p. 57.

13   "A Declaration of Concern", signed by Campbell Miller, Grady Clay, Ian L. McHarg, Charles R. Hammond, George E. Patton, and John O. Simonds, June 1966, published by the Landscape Architecture Foundation, online: https://www.lafoundation.org/who-we-are/values/declaration-of-concern, accessed 20 December 2022.

14   Gilles Clément, *Planetary Garden and Other Writings*. Philadelphia: University of Pennsylvania Press, 2015.

15   Céline Baumann and Ethel Baraona Pohl, "The Parliament of Plants and other Cautionary Tales, Where Stories Make Worlds and Worlds Make Stories", Solitude journal, vol. 1 (October 2020), pp. 72–81, online: https://issuu.com/ethel.baraona/docs/solitude_cb_ebp, accessed 2 February 2023..

16   Anabel Ford and Ronald Nigh, *El Jardín Forestal Maya: Ocho milenios de cultivo sostenible de los bosques tropicales*. San Cristóbal de Las Casas: Fray Bartolomé de las Casas, 2019, p. 47.

17   Bas Smets, quoted in Andrew Ayers, "LUMA Foundation Park by Bas Smets", *Architectural Record* (8 October 2021), online: www.architecturalrecord.com/articles/15340-luma-foundation-park-by-bas-smets, accessed 5 January 2022.

18   Sue Stuart-Smith, *The Well-Gardened Mind: The Restorative Power of Nature*, New York: Scribner, p. 10.

19   One of Brian Eno's principles for the One-Minute City. See Dan Hill, *Designing Missions: Mission-Oriented Innovation in Sweden – A practice guide by Vinnova*. Stockholm: Vinnova, 2021, online: https://www.vinnova.se/contentassets/1c94a5c2f72c-41cb9e-651827f29edc14/designing-missions.pdf?cb=20220311094952, accessed 2 February 2023, p. 434.

Paradise

The garden as a refuge and earthly paradise, a perfect world where humans and nature coexist in perfect harmony, has been a constant strand of human history since time immemorial. The garden according to this narrative is not just a physical place, but also a conceptual space onto which our hopes and dreams are projected.

The monotheistic religions celebrate the Garden of Eden as a symbol of earthly – albeit unattainable – bliss. Ancient oriental carpets were adorned with oases of superfluity, while the walled *hortus conclusus* of the Middle Ages afforded both physical safety and spiritual seclusion from the vagaries of the wilderness outside. The Japanese garden forms a bridge to the beyond, while the very formality of the perfectly manicured gardens of the French Baroque reinforced the ideal of the absolutist ruler as the seat of all power. Whether laid out on a vast scale as a utopian vision or arduously eked out from an unpromising patch of land, whether an opulent pleasure garden or an austerely rule-bound work of art, the garden as our own personal image of paradise has always reflected our own relationship with nature, and, with it, that of the society and era we inhabit.

Lucas Cranach the Elder, *The Golden Age*, c.1530

The "Golden Age" is the idea of an original ideal state, based on ancient sources and rediscovered during the Renaissance: according to the myth, this was a period characterized by perfect harmony between humans and animals and an abundance of food and freedoms. Lucas Cranach the Elder made two versions of this representation: in both, a wall protects the lush garden in which young naked people live in harmony with wild animals such as lions and deer.

A garden is a garden because it is protected from the wilderness by a fence. The word "paradise" comes from the [Iranian] Avestan word "pairí", which means "roundabout", and "daeza", which means "wall". Also, the Indo-Germanic root of the word garden, "ghorto-s", refers to a woven fence. Across history and culture, the garden is a place of tranquility and peace, where humankind wants for nothing.[1]

Nadine Olonetzky, Author

Reinagle Senr A R A pinx

London Published by D Thornton June 1805

Burke sculp

And thou, divine LINNÆUS! Deign my Reign,
O'er Trees and Plants, and Flora's beauteous Train,
Bend them obedient to my soft Controul!

Each culture and time is unmistakably expressed in gardening.[2]

Tobias Roth, Poet and essayist

↖ Thomas Burke,
*Cupid inspiring
plants with Love, in
a tropical landscape*,
c. 1805, after
Philipp Reinagle

In Roman mythology,
Cupido or Cupid
is the personification
of irresistible love.
In Thomas Burke's
colour print, Cupid
moves in a tropical
landscape in which
plants – like people –
can be joined
together as couples.

> *The garden is the smallest parcel of the world and then it is the totality of the world.*[3]

Michel Foucault, Philosopher

→ Hieronymus Bosch, *The Garden of Earthly Delights*, 1490–1500

In his triptych, Hieronymus Bosch transposes the world into a fantastic garden. While the left panel shows the original state of paradise in the Garden of Eden, the panel on the right depicts the abyss of hell. In between, people and creatures cavort, free to pursue their lustful desires, in harmony with nature.

↙ Bagh-e Shahzadeh
near Mahan, Iran, photo:
Georg Gerster, 1977

Persian gardens, as the
Bagh-e Shahzadeh,
are often lush oases in
deserts or dry areas.
So-called qanats – water
supply systems that
collect water from the
surrounding mountains
and direct its flow
into gardens – ensure
the plants are always
irrigated. The Persian
garden's division into
four areas divided by
paths and areas of water
expresses a spiritual
concept symbolic of Eden
or paradise on Earth.

↗ *The Garden Carpet*, second half of the 18th century, Iran

Traditionally, Persian rugs were treated as portable gardens that could be laid out indoors and on which people could gather. *The Garden Carpet* represents an enclosed garden with streams of water dividing the area into four sections. The rug's abstract flowers and clear geometric structure resemble a ground plan on which beds and planted areas are mapped.

In the Persian Gulf, gardens are seen as something mythical, found in the scriptures and in heaven. In our corner of the world, we have plantations. Under this canopy of tall palm trees, a complex network of aqueducts known as *falaj* (فلج) create a refuge for vegetation, provide shelter and shade, and nurture culture, faith. Life is amplified under the parasol of these trees; an oasis appears.[4]

Rashid bin Shabib, Architect and curator

Our image of Eden, most famous of all gardens, most likely derives from the rich, cultivated lower plains of Mesopotamia, and the many gardens that grew there, thousands of years ago, remain vivid cultural symbols. We can assume that the archetype of the walled garden – a re-creation of the paradise evoked in the biblical books of Genesis and the Song of Songs, and in the Epic of Gilgamesh – spread across the Fertile Crescent of the Middle East shortly after the Deluge.[5]

Christophe Girot, Landscape architect and historian

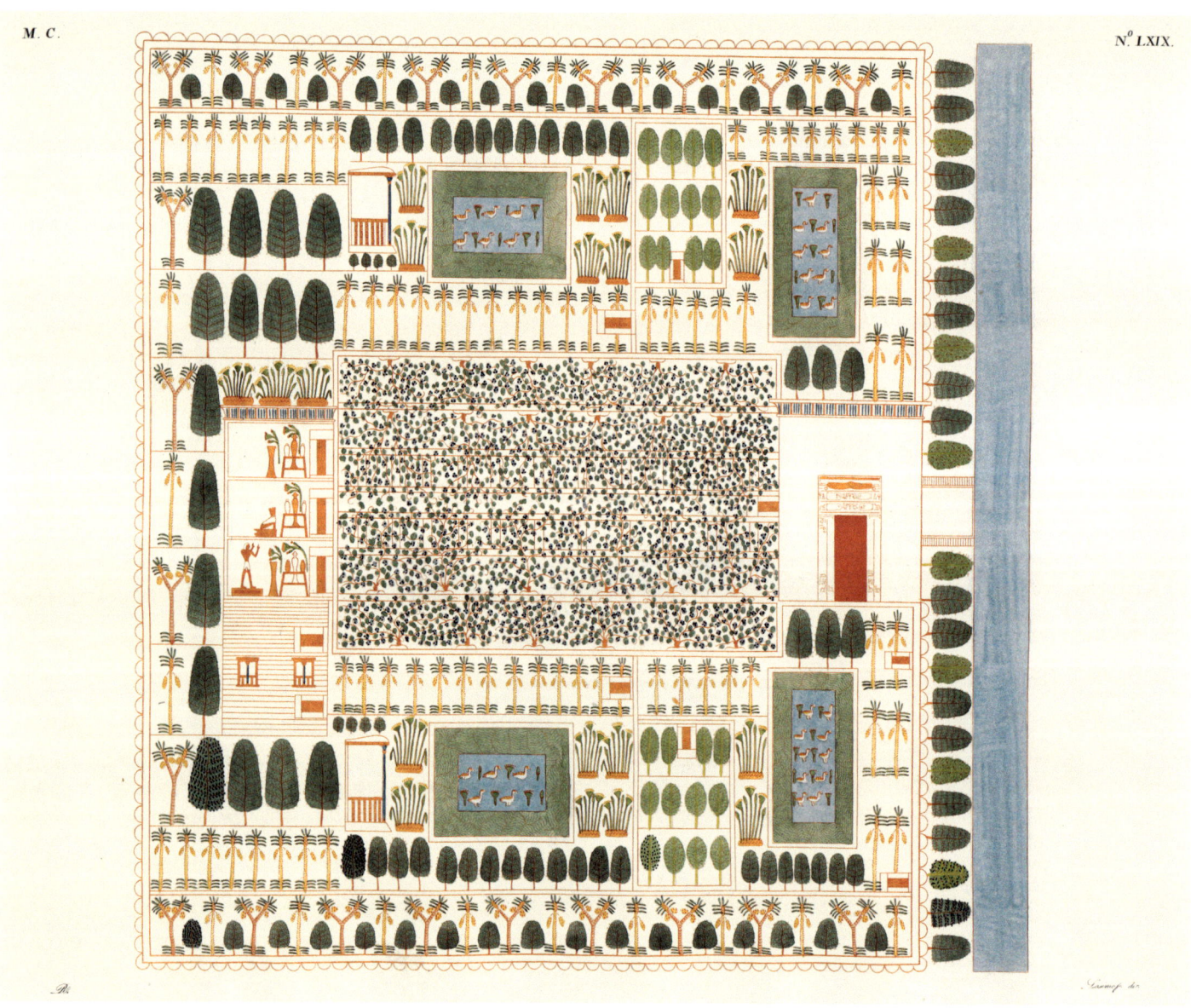

In the biblical tradition, the first human was a gardener.[6]

Annette Giesecke and Naomi Jacobs, Garden historians

← Plan-drawing by Ippolito Rosellini of Sennefer's garden found in the Tomb of the Vines near Luxor in Egypt, from *I monumenti dell'Egitto e della Nubia: Monumenti civili*, vol. 4,3, plate LXIX, 1834

A drawing made by Ippolito Rosellini in 1834 conveys an idea of how gardens were laid out in ancient Egypt. Rosellini copied a garden plan from the tomb of Sennefer, the mayor of Thebes. Depicted is the garden of the Temple of Amun at Karnak (Luxor), the largest ancient temple complex in the world.

↗ Hans Thoma, *In Paradise*, 1891

↘ Map of paradise with
various biblical scenes, from
Athanasius Kircher, *Arca Noë
in tres libros digesta*, 1675

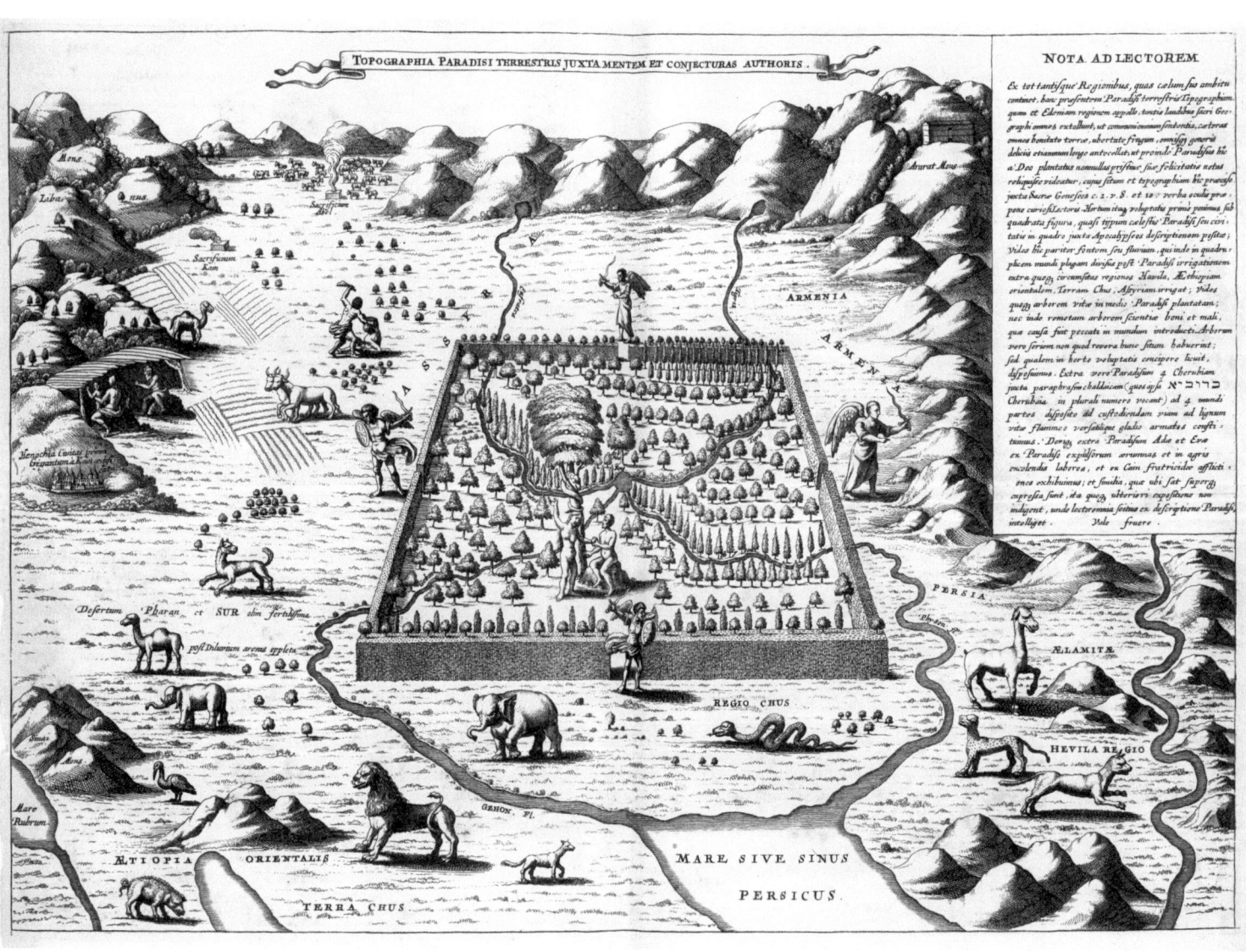

Gardens are by definition havens of peace and quiet,
order and pleasure in a chaotic and hostile world. Places
where nature is at once excluded and brought into
view [...]. The garden gathers the landscape around it
and at the same time shuts itself off from it. [...] Thus,
it is an intermediary between man and landscape.[7]

Saskia de Wit and Rob Aben, Landscape architects

↑ Upper Rhenish Master,
*The Little Garden of
Paradise*, c. 1410–20

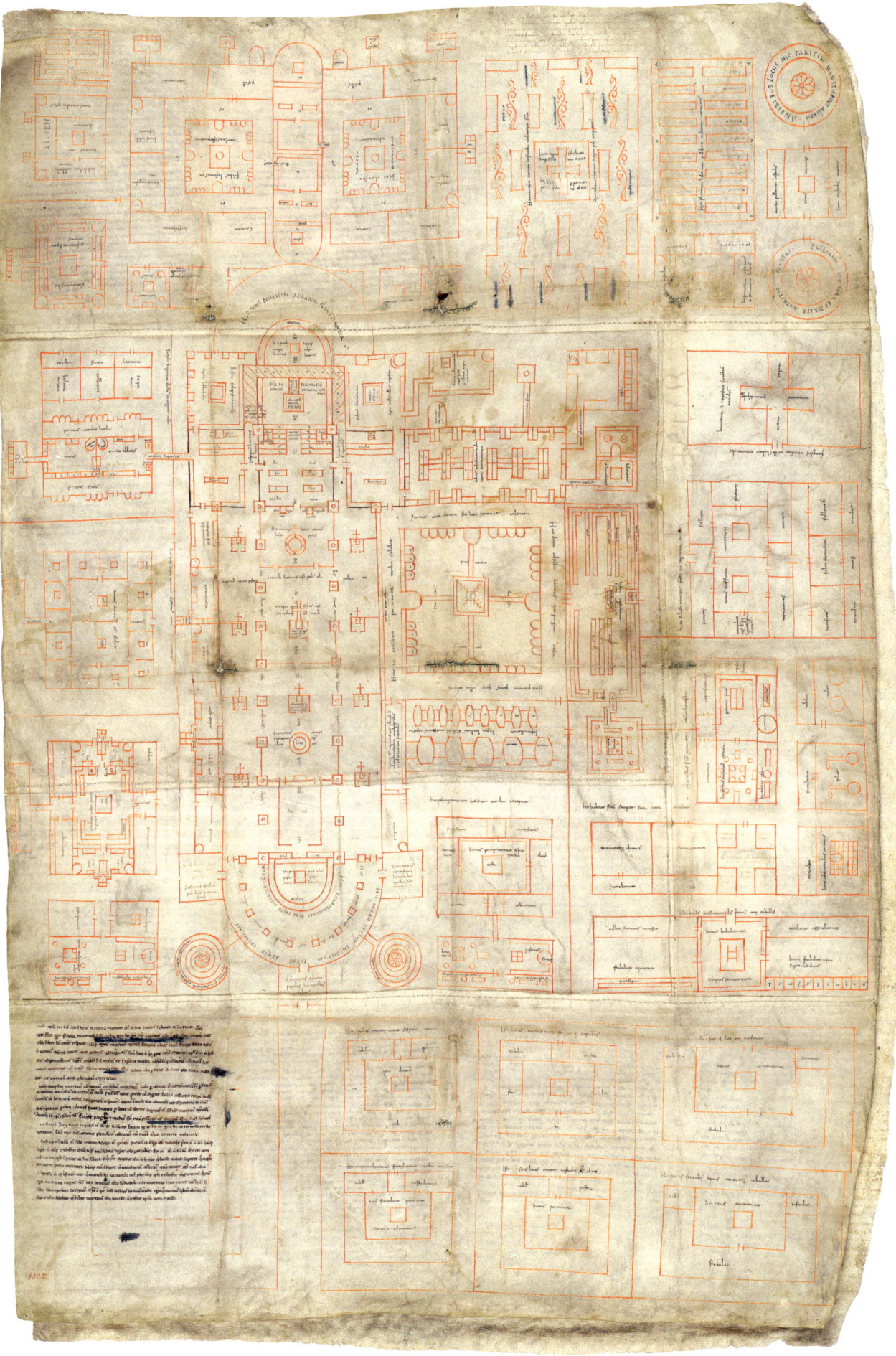

Monasteries and abbeys are symbols of the cosmic order, complete worlds in miniature. [...] The *hortus conclusus* has been described as the Ur-form of landscape architecture.[8]

Saskia de Wit and Rob Aben, Landscape architects

↗ *Iardinum domini* (Prince's garden), Medieval Botanical Garden, Palazzo Madama, Turin

→ *Hortus* (Vegetable garden), Medieval Botanical Garden, Palazzo Madama, Turin

Using historical accounts in books from 1403 to 1516, it was possible to reconstruct the moat garden around Palazzo Madama in Turin. In addition to an *iardinum domini* – prince's garden – there was a *hortus* (vegetable garden) and a *viridarium* (pleasure gardens with orchard). The walls enclosing the garden protected its crops from robberies and raids.

← Pirro Ligorio, Garden of the Villa d'Este, Tivoli, 1550–65, View from the Fountain of Neptune across the fish ponds

The park of the Villa d'Este in Tivoli near Rome is a unique example of an Italian High Renaissance garden. Cardinal Ippolito d'Este commissioned the papal architect Pirro Ligorio to construct the palazzo and gardens from 1550. The numerous water features and pools are a hydraulic masterpiece. Ligorio used the local river's hydroelectric power to create a complex system of sluices to fill and distribute water to the fountains, waterfalls, and waterways. For example, the water gushing from the Fountain of Neptune was collected in three large ponds at the foot of the fountain, in which freshwater fish were once kept. This garden not only tells of man's sovereignty over plant life, but also over water; and thus, over nature as a whole.

If people of different times and places have had very different ideas of the way a garden should be organized, it is largely because they have made very different assumptions, strongly affecting their aesthetic preferences, as to the relationship between man and nature.[9]

Elizabeth B. Kassler, Former Director of the Department of Architecture and Design, Museum of Modern Art, New York

↑ Giacomo Barozzi da Vignola, Garden of the Villa Lante, Bagnaia di Viterbo, 1486–1500

E plus que nous
auons dit ci
dessus du labou
rage du champ
dampnable des vignes
et des arbres et des Iardins
qui requierent grant art
et subtillite. Il conuient
dire apres des prez et des
bois ou il ne fault pas si
grant art ne tele paine
pour ce que le bien y vient
comme de sa nature. Et

au premier des prez pour
quoy ilz furent aez et
quel aer quel terre et si
eaue et quel place ilz veu
lent pour plus habonder
en herbe. et comme on
les fait et comment on
les procure et comment
on les renouuelle. Et du
foen qui en est le fruit
comment on le cueilt et
garde. et du prouffit qui
en vient.

The Western formal garden came from the paradise garden of Persia, to be perfected in France. From there, it became a symbol of aristocracy and authority, to many people an anathema. The rich man owns a garden; the poor man works in it.[10]

Barbara Stauffacher Solomon, Artist and landscape architect

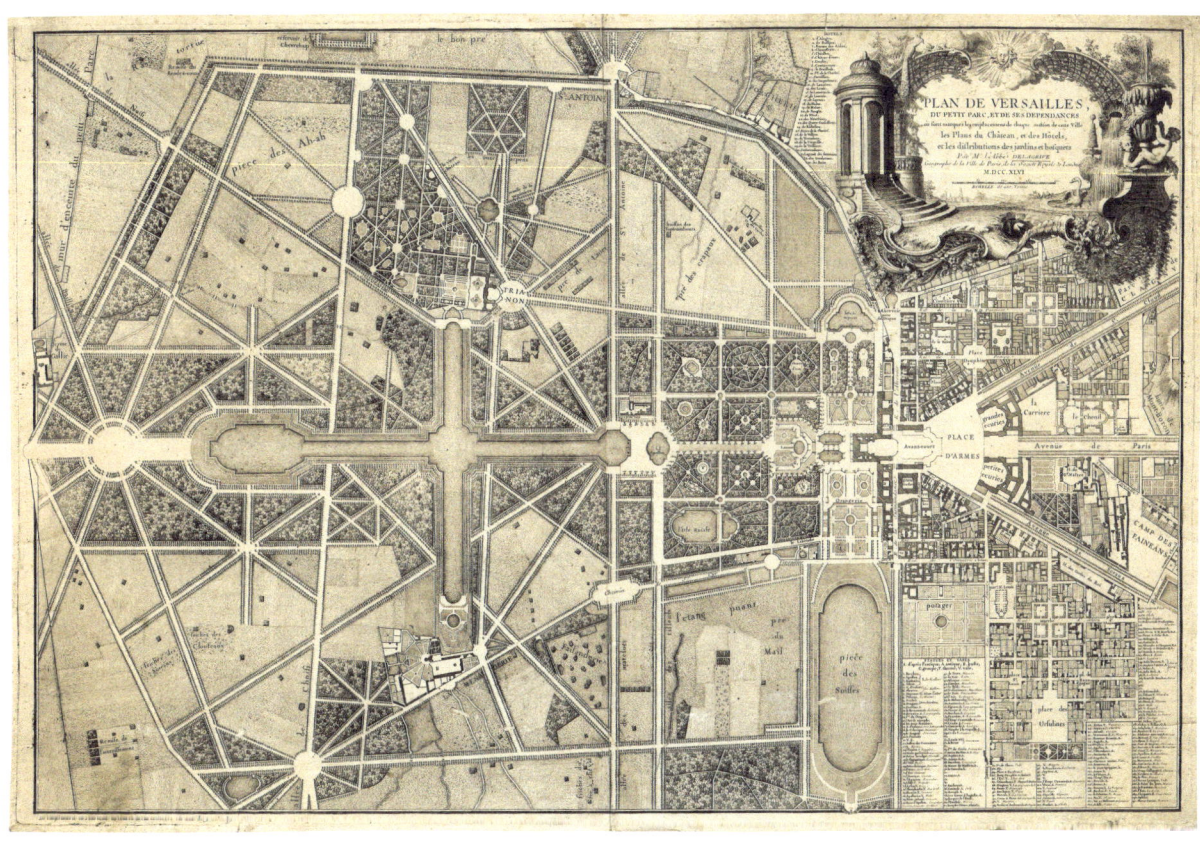

↗ Jean Delagrive, *Plan of the palace complex of Versailles*, 1746

← Illustration by Pietro de' Crescenzi from *Livre des Prouffits Champetres et Ruraulx*, fol. 198v, 1470–75

→ Jean-Baptiste de La Quintinie, Potager du roi, Versailles, 1678–83

King Louis XIV commissioned the "Potager du roi", located to the south of the Palace of Versailles, as a kitchen garden to supply the court with fruit and vegetables. The nine-hectare garden, enclosed by a wall, was divided into different growing areas and cultivated by many gardeners.

PIANTA DEL GIARDINO DEL SERENIS:
GRAN DVCA DI TOSCANA ALLA TRINI
TA DE MONTI SVL MONTE PINCIO.
Architettura di Annibale Lippi.
1 Palazzo et ingresso del Giardino uerso la Trinita de Monti.
2 Piazza e Fontana di dentro il Giardino auanti il Palazzo.
3 Galeria di Statue.
4 Piano della loggia oue sono situate le statue e li Bassirilieui antichi.
5. Obelisco con gieroglifici.
6. Mausoleo cercondato di Cipressi.
7. Tetto sotto l quale si conseruano tredici statue della fauola di Niobe.
8. Porta del Giardino nella Via Pinciana.
9. Mura antiche della Città sopra il quale e solleuato il Piano del Giardino.
10. Giardino di fiori e Agrumi.
G. Batta Falda del et inc.
G. Iac. Rossi le stampa in Roma alla pace con Priu. del S. Pont.

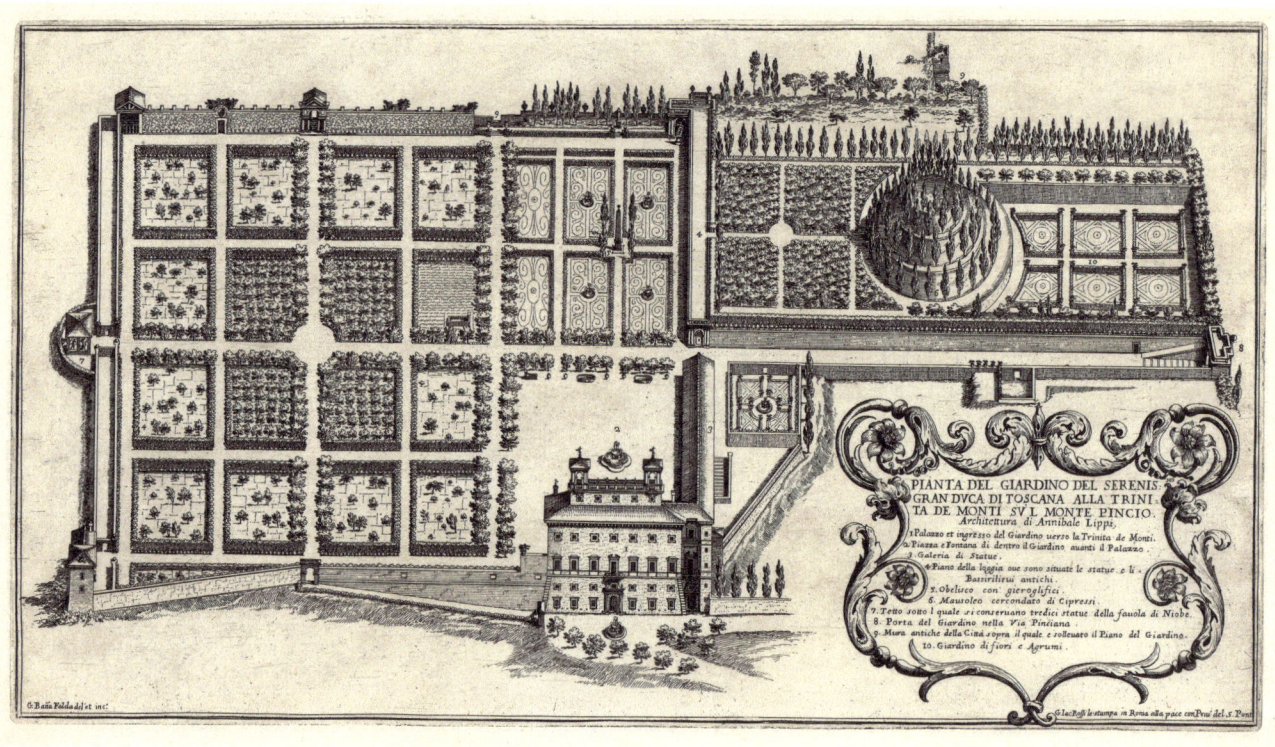

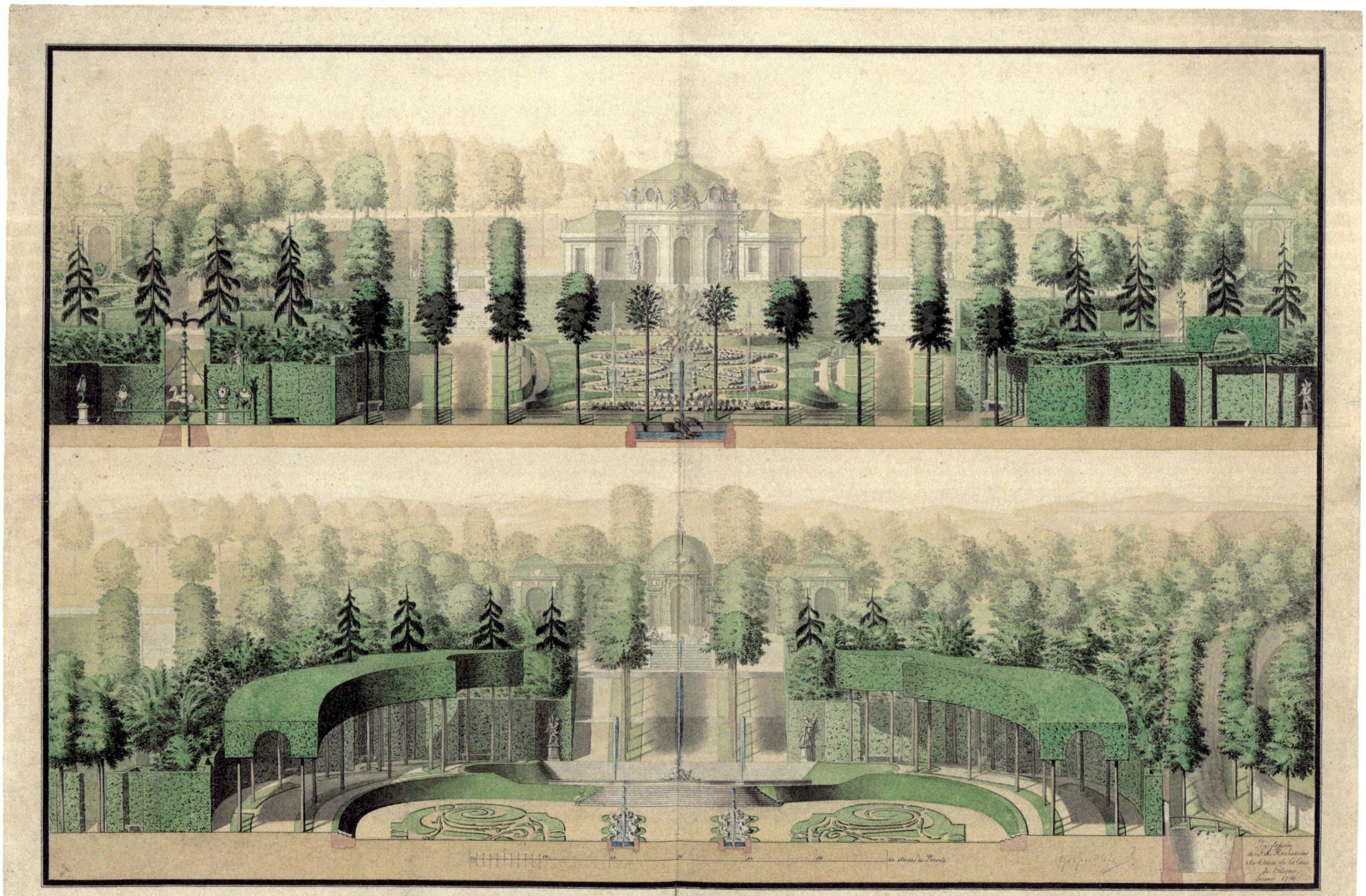

← Giovanni Battista
Falda, *Plan of the
Garden of the Villa
Medici in Rome*, 1683

↙ Design by Friedrich
August Krubsacius for
a garden at an unknown
location, 1760

↓ Julien de Cerval,
The Gardens of
Marqueyssac, Vézac,
France, designed in
the 1860s

By imposing geometric order or through miniaturization, we made [nature] appear familiar to us so that, thus constrained, it conformed to our nature or as an ideal thereof.[11]

Gerrit Komrij, Garden philosopher

It is a well-known platitude that Chinese gardens refer to the
poetic world of Chinese paintings. At the most superficial level,
we can imagine similar elements in gardens and in paintings:
pavilions, covered walkways and other built elements, plants,
rockeries, and misty scenes.[12]

Stanislaus Fung, Researcher in Chinese landscape architecture

↗ After Tang Yifen, *Aiyuan tu*
(Garden of Pleasure), 1848

Chinese depictions of gardens
and landscapes depicted
idyllic, self-contained worlds.
The medium of the hand
scroll, which was often used
for this purpose, demanded
a horizontal composition,
whereby the gardens seem to
extend infinitely in width and
the impression of vastness and
seclusion is reinforced. Varied
landscapes often consisted
of rocks, pavilions, living areas,
and waterways. Rivers and
ponds subdivide these gardens
into smaller areas.

Because artistic creation in China implies self-cultivation and self-expression, natural imagery has long been read as metaphor for the individual's values and beliefs. Flowers and plants may symbolize moral virtues; landscapes celebrating the natural order might laud the well-governed state; wilderness hermitages can suggest political isolation or protest; and gardens may be emblems of an ideal world. In every sense, therefore, images of landscape in China are reflections of both the culture and the artist's own cultivation.[13]

Maxwell K. Hearn, Curator of Chinese Painting, Metropolitan Museum of Art, New York

Above all, a garden can be an expression of the personality and world-view of its maker.[14]

Tim Richardson, Landscape historian and critic

→ Sissinghurst Castle Garden, Kent, designed by poet and gardener Vita Sackville-West and her husband Harold Nicolson in the 1930s

At Sissinghurst, nothing runs "conformably", as its co-creator Vita Sackville-West put it. Playing with the modernist literary form of stream-of-consciousness, the design makes virtues of illogicality, disorientation, fragmentation and illusion. Sissinghurst has become an inspiration for garden-makers.[15]

If what the gods made is nature, then the garden
is the part that the gods forgot to make. So it
is up to us to take the place of the gods and make
gardens. We ourselves must become the gods.[16]

Mirei Shigemori, Landscape architect and historian

↗ Mirei Shigemori, Garden of the Maegaki family, Higashi-Hiroshima, 1955. The wave becomes a new linear element.

← Mirei Shigemori, Ryôgin-an temple garden, Kyoto, 1964. The surface acquires colour and a new meaning.

→ Mirei Shigemori, Tōfuku-ji Hôjô, Kyoto, 1939. The points of the stone settings symbolize the residences of the gods.

Mirei Shigemori experienced the Japanese garden as a cultural asset trapped in its thousand-year history. Throughout his life, he dedicated himself with great verve to its renewal and conceived the future of the *karesansui* garden, the ancient concept of the dry landscape garden, thus the very root of garden art in Japan. As a painter, he saw the dots, lines and planes in gardens with different eyes than his fellow gardeners. His dramatic stone settings, playful waves and colourful gravel surfaces are therefore his multilayered answer to the question of the future of the Japanese garden. Just as at the very beginning of the Japanese garden, stones still represent the places for the gods in Shigemori's gardens.[17]

GABRIEL GUÉVRÉKIAN
Jardin (maquette). à Hyères

Les Arts de la Maison
Hiver 1926

XXVII

Éditions Albert Morancé

Tradition excluded from the gardened
area all those living species, both
animal and vegetable, which escaped
the gardener's control.[18]

Gilles Clément, *Gardener and garden philosopher*

← Gabriel Guévrékian, Architectural model of the triangular garden of the Villa Noailles, Hyères, from *Les Arts de la Maison*, Winter 1926, plate XXVII

→ Gabriel Guévrékian, Garden of the Villa Noailles, Hyères, designed in 1926

In the early 1920s in France, a number of gardens were created that had a significant influence on modernist landscape architecture. A radical example is Gabriel Guévrékian's chessboard garden, which he designed in 1926 for the cubist Villa Noailles in Hyères. While the Vicomtes de Noailles commissioned the architect Robert Mallet-Stevens to build the villa, Guévrékian was to develop a cubist garden. Guévrékian designed a composition of terraces, some planted and others paved with concrete and mosaics. These rectangular slabs were arranged on a tapering, triangular surface surrounded by a wall. Irregularly shaped beds were placed along the sloping verges. With this unprecedented garden design, Guévrékian combined cubist forms with new materials, creating a landscape architecture that was closely based on the ideals of modern art.

← Luis Barragán, Jardines del Pedregal, Mexico City, 1954–52, show garden with landscape and enclosing walls, photo: Armando Salas Portugal

In the late 1940s, the Mexican architect Luis Barragán conceived plans for a vast residential subdivision on the site of a lava field south of Mexico City. His visionary concept for the Jardines del Pedregal aimed to merge the Indigenous landscape with architecture and landscaping based on a modern aesthetic. The internationally acclaimed project responded to the needs and expectations of the rising Mexican middle class and reflected the spirit of a budding national identity.

↘ Francisco Artigas, Casa Federico Gómez in Jardines del Pedregal, Mexico City, 1952

↓ Aerial view of Jardines del Pedregal, Mexico City, 1952

CIA. MEXICANA AEROFOTO, S. A.

9273

[...] we have found that, in order not to harm and spoil this landscape beauty, and to create beautiful architectural forms that will not compete with them, they must be of such simplicity – abstract in quality, preferably straight lines, flat surfaces, and primary geometric forms.[19]

Luis Barragán, Architect

↑ Alvar Aalto, Villa Mairea,
Noormarkku, Finland,
1938–39

In 1939, Finnish architect
Alvar Aalto built a two-storey,
L-shaped building for Maire
and Harry Gullichsen, framing
a pool with a courtyard. This
large, organically shaped
pool is composed of rounded
tiles and curved paving
slabs and is thought to be
one of the first non-linear
pools. Embedded in a wide
lawn and situated at the
edge of the forest, the
pool was designed to appear
like a natural lake in the
Finnish landscape.

The organic nature of modern building demands that architecture does not stop with the house, but continues out into the landscape.[20]

Christopher Tunnard, Landscape architect and writer

The objective is to shift the premises for urban lifestyle and access to nature in tall buildings away from formulaic Modernist design. This alternative offers inhabitants a unique opportunity to assert their personal presence and celebrate their choices of vegetation in the cityscape.[21]

James Wines, Artist

← James Wines, *Highrise of Homes*, 1981

With his speculative and some realized projects, American artist James Wines – with his design firm, SITE – was one of the most important protagonists of the rising interest in architectural environmentalism in the 1970s and 1980s. One of his most distinctive motifs is the "Highrise of Homes", which began as a research project on housing. The drawing combines the typology of the high-rise with the American Dream of the suburban single-family home. Thus, Wines addresses the fact that while vertical, cellular forms of housing bring with them undisputed social and ecological advantages, this rational view nonetheless will always be in conflict with the desire for an individualized living environment that features a private garden.

↗ Residents gardening in the Bärensteinstraße, Berlin, photo: Peter Zimmermann, 1982

Gardening in the GDR was much more than a hobby – gardening offered opportunities for socializing and self-sufficiency. Residents planned, planted, cultivated, and managed parcels of land in allotment gardens on the outskirts of the city, and utilized the green areas amidst housing estates for the same purpose.

If the planet is a garden, we are all gardeners – perhaps we are not aware of it, yet our choices and lifestyles have an impact on the biosphere and on our collective vital space.[22]

Gilles Clément, Gardener and garden philosopher

1 Edited from: Nadine Olonetzky, *Inspirations: A Time Travel through Garden History*. Basel: Birkhäuser, 2017, p. 18. (Correspondence between Nadine Olonetzky and Viviane Stappmanns, 14.10.2022.)

2 Tobias Roth, *Gärtnern mit Sprengstoff (1912–1929). Eine Collage*. Die Grüne Reihe, vol. 2, Berlin: SUKultur, 2022, p. 27.

3 Michel Foucault, "Des espaces autres. Hétérotopies", *Architecture /Mouvement/ Continuité*, no. 5, Paris 1984, pp. 46–49. (The text was the basis of a lecture given by Michel Foucault in March 1967.) English: Michel Foucault, "Of Other Spaces Heterotopias", trans. from *Architecture /Mouvement/ Continuité*, no. 5, 1984, pp. 46–49, online: https://foucault.info/documents/heterotopia/foucault.heteroTopia.en

4 Correspondence between Rashid bin Shabib and Carolina Maddè, 10.11.2022.

5 Christophe Girot, *The Course of Landscape Architecture: A History of our Designs on the Natural World, from Prehistory to the Present*. London: Thames & Hudson, 2016, p. 30.

6 Annette Giesecke and Naomi Jacobs, *The Good Gardener*. London: Artifice Books on Architecture, 2015, p. 16.

7 Rob Aben and Saskia de Wit, *The Enclosed Garden: History and Development of the Hortus Conclusus and its Reintroduction into the Present-day Urban Landscape*. Rotterdam: 010 Publishers, 1999, p. 49, p. 10.

8 Ibid., p. 10, p. 5.

9 Elizabeth B. Kassler, *Modern Gardens and the Landscape*. New York: The Museum of Modern Art, 1964, p. 5.

10 Barbara Stauffacher Solomon, *Green Architecture and the Agrarian Garden*. New York: Rizzoli International Publications, 1988, p. 57.

11 Quoted in: Laurie Cluitmans (ed.), *On the Necessity of Gardening: An ABC of Art, Botany and Cultivation*. Amsterdam: Valiz, 2021, p. 13. (Originally from: Gerrit Komrij, *Over de noodzaak van tuinieren*. Amsterdam: Bert Bakker, 1991, p. 61).

12 Stanislaus Fung, "Memory, Direct Experience, and Expectation: The Contemporary and the Chinese Landscape", Christophe Girot and Dora Imhof (eds), *Thinking the Contemporary Landscape*. Princeton, NJ: Princeton Architectural Press, 2017, p. 249.

13 Maxwell K. Hearn, *Cultivated Landscapes: Chinese Paintings from the Collection of Marie-Hélène and Guy Weill*. New Haven: Yale University Press, 2002, p. 4.

14 Correspondence between Tim Richardson and Carolina Maddè, 13.01.2023.

15 Ibid.

16 Mirei Shigemori, "Shinsakuteiki", *Shigemori Mirei Sakuhinshû: Niwa – Kamigami eno Apurôchi* (Shigemori Mirei's Collected Works: Gardens – Approach to Gods). Tokyo: Seibundô Shinkô Sha, 1976, p. 287.

17 Correspondence between Christian Tschumi and Carolina Maddè, 23.01.2023.

18 Gilles Clément, "In Practice: Gilles Clément on the Planetary Garden", *The Architectural Review* (16 February 2021), online: https://www.architectural-review.com/essays/in-practice/in-practice-gilles-clement-on-the-planetary-garden

19 Luis Barragán, "Gardens for Environment, Jardines del Pedregal", *Journal of the American Institute of Architects*, no. 4 (April 1952), p. 171.

20 Christopher Tunnard, "Modern Gardens for Modern Houses: Reflections on Current Trends in Landscape Design", *Landscape Architecture Magazine*, vol. 32, no. 2 (January 1942), published by the American Society of Landscape Architects, pp. 57–64.

21 Correspondence between James Wines and Viviane Stappmanns, 18.01.2023.

22 Gilles Clément, "In Practice: Gilles Clément on the Planetary Garden", *The Architectural Review* (16 February 2021), online: https://www.architectural-review.com/essays/in-practice/in-practice-gilles-clement-on-the-planetary-garden

Garden
Politics

The idyll is deceptive, however, as the ideal garden has long been shaped by influences both political and commercial. Exactly whose interests are manifested in gardens becomes apparent only on closer inspection of the objects and milestones of garden history, which tells us that gardens have always been less of a refuge and more of a battlefield than we perhaps thought.

Who owns a garden, the purpose it is to serve, and the place it is to occupy in the urban environment – these are questions to which numerous concepts of urban and regional planning have sought to give answers. Among the best known of these are the "garden cities" of the early twentieth century. Even the act of gardening has been instrumentalized for political purposes. National governments in times of crisis have declared growing vegetables a patriotic duty, while citizens with grievances have often wielded spades to lend emphasis to their demand for a say or simply a full stomach. Many an exotic flowerbed has deep roots in the history of colonialism, and how the dream of a luxuriant but easy-to-maintain recreational paradise should look is determined not just by gardeners, but also by the designers who stage its colourful displays.

Dreaming of a Global Garden

by Luke Keogh

Today, we are accustomed to our gardens being places filled with roses, jasmine, fuchsia, chrysanthemums, and rhododendrons, yet many of the most common varieties originate from faraway countries. This development is owed in large part to the Wardian case: a box made of wood and glass, essentially a movable terrarium, created by London surgeon and amateur naturalist Nathaniel Bagshaw Ward (1791–1868). The invention came at the height of global colonial expansion, so initially there was a widespread enthusiasm about a technology that could move live plants successfully. Over the following century, the Wardian case contributed to the transformation of the global environment.

In 1829, Nathaniel Ward placed soil, dried leaves, and the pupa of a sphinx moth into a sealed glass bottle, intending to observe the moth hatch. But after fern and meadow grass sprouted from the soil, he discovered that plants enclosed in glass containers could survive for long periods without watering. In his London home, Ward then experimented with a variety of plants and realized that the cases could also be used to transport plants. Ward created sturdy timber cases and decided to test how far they could travel by taking the longest sea journey then known – to Australia. The cases' sloping roofs held glass inserts, creating a microenvironment that allowed plants placed in pots or soil and secured with battens to survive. The cases required little care and no fresh water on the long sea voyage. The first test run in 1833 from London to Sydney with ferns, mosses, and grasses proved to be successful.

↗ Wardian case used by the Royal Botanic Gardens, Kew, c. 1870

The impulse to move plants across the oceans is as old as the human passion for travel. Ward was not the first to move plants around the world, but the idea of the enclosed system was his discovery. With the colonial project of many European leaders at its height, his invention came at a time of great demand. Because of Ward's determination as an experimenter and his far-flung friendships and connections to many of the leading scientists and horticulturalists of the day, his technology was quickly adopted and widely used. As early as 1835, Ward's friend George Loddiges, an influential nursery businessman and scientist, put into use nearly 500 cases around the globe. Initially, the glazed cases became the preferred method of moving plants for commercial nursery firms, who moved anything from ferns, cedars, orchids, and even orange trumpet creepers to useful plants, such as bananas, custard apples, and sago palm. Later, the world's leading botanical institutions began using the case to extend their collections and distribute plants between overseas colonies, including the Royal Botanic Gardens, Kew; Jardin des Plantes, Paris; and the Berlin Botanical Gardens.

↖ Wardian cases in preparation for leaving the Royal Botanic Gardens, Kew, c. 1940

Even though transporting plants in Wardian cases was more successful than using other methods, they still required expert packing and professional care during the journey. The hard work of moving plants was done not only by horticulturalists, botanists, and gardeners, but also by colonized peoples, indentured workers, and slaves. It was also the intimate knowledge of Indigenous peoples about local plants that allowed plant hunters and foreigners to locate and exploit certain plants – the high-yielding cinchona plants from South America is a classic example here. In addition to Indigenous labour, the colonial powers also appropriated their lands for the global plantation economy. The Wardian case was of special importance in establishing plantations by transporting key species, such as avocados, bananas, cacao, coffee, cinchona, gutta-percha (to make resin for underwater cables), fruit trees, timber trees, mangoes, tea, and, above all, rubber. The source of the purest and most elastic latex was Pará rubber (*Hevea brasiliensis*), native to Brazil and transplanted to different parts of Asia.

↘ Wardian cases used for the shipment of plants from the Botanic Garden in Buitenzorg (now Bogor), West Java, 1913

Tab. XII.

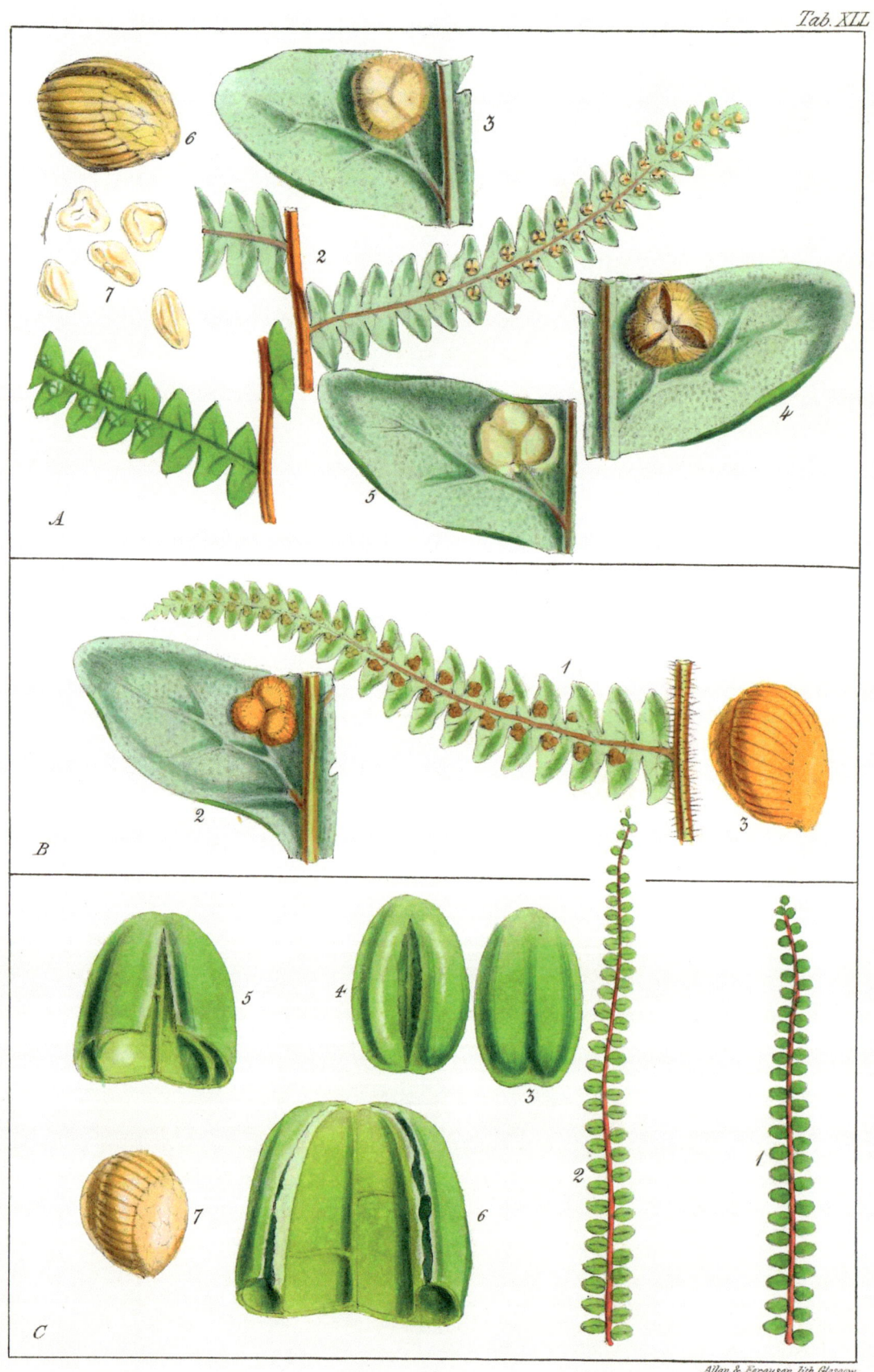

The first plant that was transported from Australia to London in 1834 was a delicate coral fern – *Gleichenia microphylla*. This exact individual would later be drawn by Franz Bauer and find its way into William Hooker's *Genera Filicum* (1842). From the 1850s onwards, ferns were all the rage in Victorian Britain. The so-called fern fever or Pteridomania was closely connected to the Wardian case. In one of his early experiments Ward had used a fern, which had stunted in the polluted London air of his garden but thrived when placed in an enclosed glass bottle. Following the bottle experiment, Ward switched to small movable glass cases for the plants held in his house, thus starting a fashion that eventually spread throughout Britain, Europe, and the United States. Ferns displayed in a parlour-style Wardian case epitomized refinement, and as an indoor adornment completed the most tasteful living space.

← *Gleichenia microphylla*, illustration of fern by Franz Bauer, from William Jackson Hooker, *Genera Filicum*, 1842

Magnolia Purpurea.

Gardening was a wildly popular pastime in the nineteenth century. This was not limited to the upper class; the emerging middle classes were also turning enthusiastically to gardening. With the spread of colonial empires and (in many cases) colonial settlers, this new passion also spread across the globe, with many nurseries specializing in exotic plants. Ornamental plants that arrived in Europe included shrubs of the Magnolia genus, such as *Magnolia liliiflora* and *Magnolia stellata* from Japan, or *Lonicera japonica*, the Japanese honeysuckle, which turned out to be an invasive plant that today is found worldwide.

↗ Purple magnolia (*Magnolia liliiflora*), illustration from Mary Ann Burnett, *Plantae Utiliores; or Illustrations of Useful Plants, Employed in the Arts and Medicine*, vol. 4, 1850

At the same time as becoming a transport technology, ornamental Wardian cases in various shapes and sizes became indoor features, creating favourable climatic conditions for the fragile and exotic plants purchased at specialist nurseries. The box that held the soil was usually made of well-aged mahogany, and it had a number of supports for the glass cover which featured a small door. Over the years, this and other versions became a prized adornment of homes throughout Britain, Europe, and the United States.

→ Illustration of an indoor Wardian case, from Henry T. Williams, *Window Gardening: Devoted Specially to the Culture of Flowers and Ornamental Plants, for Indoor Use and Parlor Decoration*, 1874

Besides being an accomplice to colonial exploitation, the Wardian case became an accessory to a whole range of complex ecological transformations. Other than plants, an average-sized case also contained more than 55 litres of soil, which unintentionally transported many other species such as fungi, algae, nematodes, insects, mites, and pathogens. Some of these, but also many of the plants, became a serious threat to the ecosystems in which they were introduced. Ironically, the Wardian case was regarded as both culprit and cure. For example, insects were moved around the globe to feed upon pests or out-of-control invasive plants; sometimes this was highly successful, such as the transportation of the cactoblastis moth to Australia to stop the prickly pear outbreak in the 1920s, but more often biological control had disastrous consequences like the extinction of the coconut moth on Fiji. In the 1940s, most countries banned the importation of soil and the Wardian case was phased out.

↓ Wardian cases containing the wasp *Encarsia lahorensis* for the control of the citrus white fly, leaving Lahore for Orlando, Florida, 1911, from Russell Woglum, *Report of a Trip to India and the Orient in Search of the Natural Enemies of the Citrus White Fly*, 1913

Ward used every opportunity to promote his case. A major milestone in its acceptance was the Great Exhibition of the Works of Industry of All Nations held in the Crystal Palace, London, in 1851. The Wardian case was featured as an industrial innovation. It appeared in various parts of the exhibition and in different versions, as adorned parlour cases and as plain travelling cases. The Great Exhibition was an important location to emphasize the mission of colonialism, and a significant design element of the Crystal Palace was the enormous amount of glass: more than 300,000 sheets of glass were used in its construction. It inspired the cheap manufacture of glass in Britain, which led to a rapid uptake of gardening under glass, not only in Britain but also in Europe and in the colonies. Despite the commercial success of the Wardian case, its inventor died destitute in 1868. Ward had never patented his invention because he believed that it should benefit everyone.

← Interior view of Joseph Paxton's Crystal Palace in London, from *Dickinson's Comprehensive Pictures of the Great Exhibition of 1851*, 1854

Luke Keogh's research into the Wardian Case first appeared in his book *The Wardian Case: How a Simple Box Moved Plants and Changed the World*, Chicago: University of Chicago Press, 2020.

From Garden City to Post-Industrial Landscape

Gardens have always reflected an imaginary future, even acting as utopian repositories for a yearned-for relationship to nature. To the extent that gardens represent social ideals as well as ideological systems, they can also be understood as political sites in themselves, as OLIVER SUKROW demonstrates in the following case studies.

curving eroded hillock - harder terrain

plantings of Bishop
pines in the draws
cool, moist
protected.

Beach
& shingle
Talus- broken
away from
cliffs edge

seal rocks @
low tide

follows force
of waves

former cliff's
edge - now
free-standing rock

S/R. looking Southward - the
Shapes & processes of the Coast
Oct 24 - '76

Halprin

It was during the Enlightenment, around 1800, that the ostensibly "natural" English landscape garden became the new model to aspire to – not just in terms of aesthetics and design, but also morally. The "garden revolution" that first showed shoots in the United Kingdom had however eclectic roots. The natural looking garden conveyed an understanding of freedom that was diametrically opposed to the courtly, hierarchical systems presided over by the absolutist rulers of the baroque period, with all their attendant limitations. In his multivolume *Theorie der Gartenkunst* (1775–85), garden theorist Christian Cay Lorenz Hirschfeld took up the philosophical ideas and in particular the criticism of civilization of French Enlightenment thinker Jean-Jacques Rousseau, in which the reclamation of an idealized (if hypothetical) state of nature played a key role.

The process of defortification, which entailed tearing down medieval walls and gates of cities, gave rise to new urban spaces for sociality and activities that the upwardly mobile middle classes soon claimed as their own. In cities like Vienna, the old bastions and city walls were turned into promenades, tree-lined boulevards, and public green spaces. This also helped improve sanitary conditions, and to this day serves as a graphic reminder of how the city literally burst out of its medieval confines. Various urban planning measures such as the zoning of residential areas with parks and boulevards, and the more fluid because no longer fortified boundaries between town and country, projected an image of the city that was no longer turned in on itself but open and outward-looking. Monuments, vistas, and thoroughfares served to dramatize this same process of opening up. When urban planner Christian Zais began expanding the city of Wiesbaden in 1818, for example, he proceeded from each side of the pentagon defining the limits of the

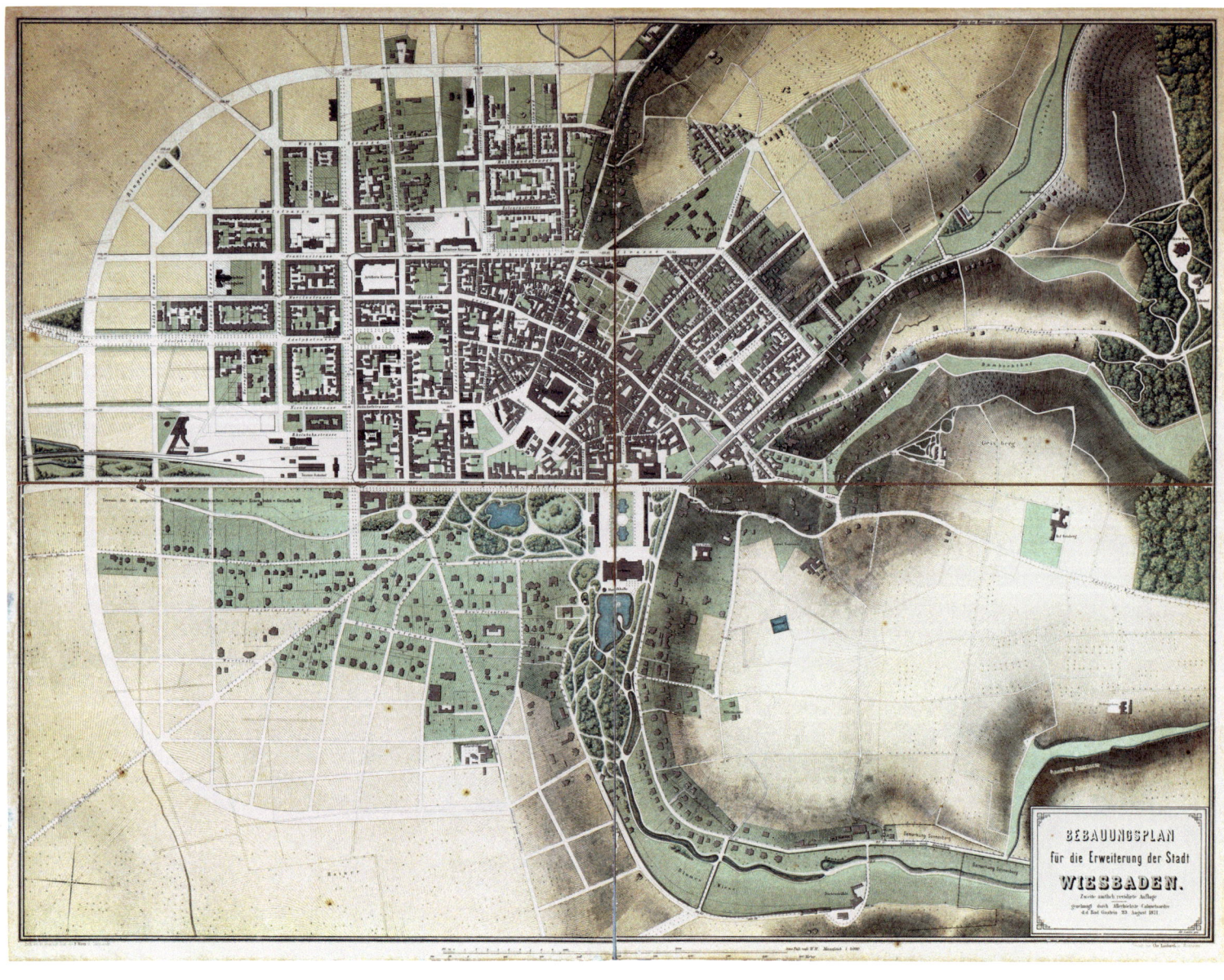

Green spaces helped improve sanitary conditions, and to this day serve as a graphic reminder of how the city literally burst out of its medieval confines.

↗ Richard Riemerschmid, Terraced houses in the garden city of Hellerau near Dresden, c. 1910

← Alexander Fach, Development plan for the expansion of the city of Wiesbaden, 1871

medieval town centre and worked outwards. As the nineteenth century progressed, the former residence of the Dukes of Nassau was enlarged and "prettified", and more and more baths and pump rooms built to turn it into a major spa town. The new, whitewashed facades and neoclassical buildings lining the promenades "looked out" onto the surrounding countryside – one vast landscape garden, as it were.

The city planning and sanitation reforms initiated in the early days of the nineteenth century had to be adapted to keep up with the greatly changed socio-economic conditions just five decades later, a time when gardens and parks were assigned an important and often politically relevant role. Industrialization saw large numbers of people flocking to the cities and

more often having to live in subhuman conditions. Their squalid homes, compounded by long working hours in exchange for a mere pittance, led to unrest and uprisings. The people's demands for better living and working conditions were taken up by mostly left-wing political parties and movements, who made them a matter of some urgency. Cities and municipalities, usually acting in concert with local employers, responded by implementing social housing projects that also incorporated green spaces. Industrialists with a social conscience followed suit, sometimes building whole estates for their workers.

The first to float the idea of a garden city – a concept that continues to resonate today – was almost certainly the British stenographer and social reformer

Ebenezer Howard, whose influential book *Garden Cities of To-morrow* was published in 1902; an earlier version *To-morrow: A Peaceful Path to Real Reform* having previously appeared in 1898. The Deutsche Gartenstadtgesellschaft (German Garden City Society) was also founded in 1898, in Berlin. Howard, who did not consider himself an "urbanist" as such, produced vivid diagrams that sketched out the garden city's potential as a place where the specific advantages of both urban and rural life might coalesce. With their meandering streets, picturesque squares, small-scale development of detached and semi-detached houses, all with a front and back garden, Germany's first garden cities such as Georg Metzendorf's Margarethenhöhe in Essen (built from 1906) and Richard Riemerschmid's Gartenstadt Hellerau in Dresden (from 1909) sought to evoke the village of the pre-industrial era. Green spaces for communal use – supplemented by vegetable patches of the residents – played an important role in these developments, just as they would in the social housing of the twentieth century. As much as the romanticized ideal of small-town life might be criticized, including by Friedrich Engels in his pamphlet "The Housing Question" of 1872, the garden city nevertheless offered an attractive alternative to urban megastructures. These reform projects were an important milestone in the development of the modern dwelling and the modern way of life, some aspects of which were indeed utopian and aimed at bringing about social change.

The modernists would also take up the idea of the garden city in their social housing developments of the interwar period. While they radically simplified the architectural language of forms, they left the idea of a vegetable patch for every dwelling virtually unchanged. The gardens of the New Frankfurt developments of Praunheim (built 1926–29) and Römerstadt (1927–29) were the work of urban planner Ernst May, in collaboration with the director of horticulture Max Bromme and the landscape architect Leberecht Migge. The latter insisted that every new subsistence level housing unit should have its own garden for growing fruit and vegetables. Besides easing the financial burden of its occupants in times of crisis, such gardens would

While the modernists radically simplified the architectural language of forms, they left the idea of a vegetable patch for every dwelling virtually unchanged.

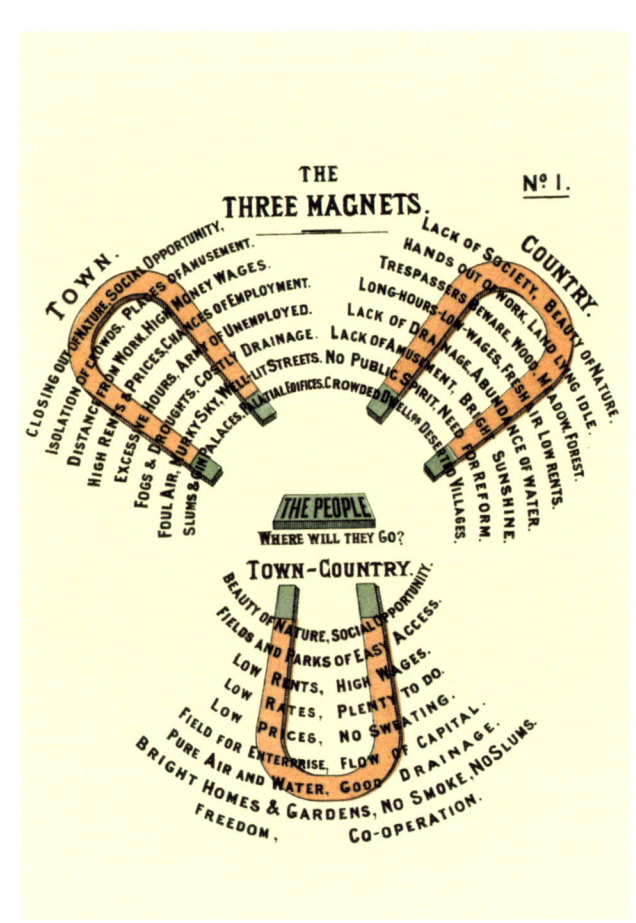

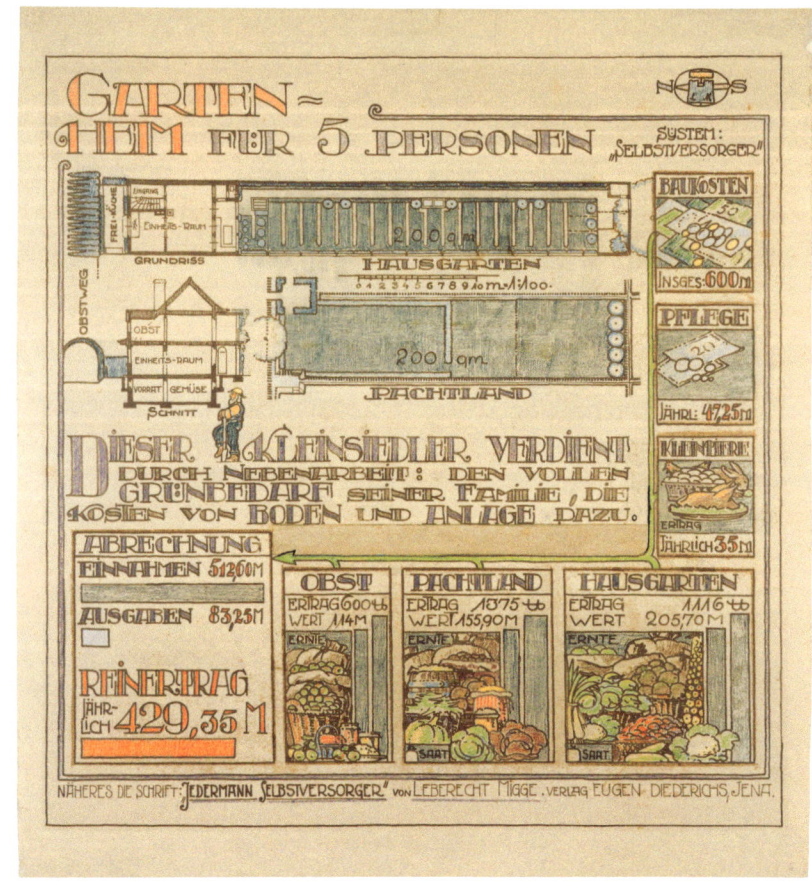

↑ Leberecht Migge, Allotment gardens in Siedlung Römerstadt, Frankfurt am Main, 1928

↖ A greener city, illustrated by a diagram of the Three Magnets (Town, Country, and Town-Country) in Ebenezer Howard's book *Garden Cities of To-morrow*, 1902

← Leberecht Migge, Garden home for five people, system "self-sufficient", 1918

also give them salutary hands-on contact with nature and the soil. Migge presented his concept of "green colonization" in an appraisal report called "Grünpolitik der Stadt Frankfurt am Main" (The Greening of Frankfurt am Main), published in 1929. His demand for "less green welfare, more green self-help" was to be achieved by increasing the amount of land available for "productive green spaces" as opposed to parks, so that local residents could grow their own food. However, Migge did not want the future planning of house and garden to be left to the occupants only; he clearly felt that the design of his "consummately engineered gardens" could no more be entrusted to amateurs than that of the houses. The regional plan that Migge drew up for the New Frankfurt development was ambitious, providing not just residential areas but farm plots and garden zones. Everything was geared to

a combination of economy and ecology. Migge's concept for the comprehensive greening of Frankfurt am Main, including a green belt of allotment gardens, would only be repeated on a comparable scale in Germany many years later – the renaissance of the allotment garden in West Germany during the 1970s was born out of an anti-capitalist sentiment and the nascent environmentalism of that era, whereas growing your own fruit and vegetables had long been an established part of life in East Germany.

Exhibitions were an important forum for the presentation and discussion of contemporary garden projects. Not only did they provide an opportunity for (international) comparisons, but they also revealed the larger political and socio-economic contexts and debates to which the concepts presented belonged. Whether held under the auspices of a World's Fair,

Entwurf: Professor Max Läuger in Karlsruhe.

Exhibitions were an important forum for the presentation and discussion of contemporary garden projects.

trade fair, or industry show, or staged as events dedicated to gardening and garden politics in their own right, garden shows were an especially important vehicle for ideas in the years immediately preceding World War I. Alone the architectural typology of the Crystal Palace that gardener Joseph Paxton designed for the first ever World's Fair in London, the Great Exhibition of 1851 (p. 62) was proof of the close ties between gardening, engineering, and architecture; Paxton having drawn inspiration for his glass and steel structure from the ribbed underside of the giant lily pads of the *Victoria amazonica* water lily. His groundbreaking design even integrated some of Hyde Park's mature trees, which were simply housed for the duration. The Crystal Palace was thus hothouse, plant show, and exhibition hall rolled into one. Paxton's "green" exhibition architecture presented a model that would reverberate well into the twentieth century.

The principles of the quasi-architectural (private) garden were developed in Darmstadt, Mannheim, and Vienna around 1900. These strictly geometric gardens tell of the pre-eminence of architectural forms in the gardens of the modernists, who tended increasingly to prefer "artificiality" to the (ostensible) "naturalness" of the English landscape garden. Rationality, rigour, restraint, and order not only left their stamp on the architecture of the World War I period, but also influenced the development of the garden as a political echo chamber.

The "geometrization" of the garden around 1900, coupled with the ever-increasing influence of architects on garden design, had an impact not only on private gardens but on the design of public green spaces too. The new discipline of urban planning became a driver of innovation in the planning of communal parks, gardens, and green spaces, which all

↑ Max Laeuger, Site plan of the rose garden for the Internationale Kunst- und Gartenbauausstellung, Mannheim, 1907

→ Moshe Safdie, Habitat 67, opened during the 1967 International and Universal Exposition, Montreal

became vital to urban growth and building regulation strategies. The same period saw greater discussion of functionalist concepts, such as greenways or green wedges as part of the urban planning toolkit. The purpose of these green spaces was to increase fresh air, break the monotony of urban development, and satisfy the need for places to socialize and engage in sports and recreation. The dawn of the twentieth century saw the public gardens championed by Enlightenment garden theorists such as Hirschfeld transform into a new typology, that of the multifunctional public park. The directors of buildings and city architects (leading city officials, in other words) endowed the new suburban peripheries of expanding, sprawling cities with generous green spaces. This was the subject of "Das sanitäre Grün der Städte" (The Sanitary Green of the Cities), Martin Wagner's dissertation of 1915, in which the man who would later become Berlin's chief city planner outlined the importance of the public park to twentieth-century city dwellers.

The green spaces created in the years 1933–45 (and in East Germany after 1949) took their cue from early twentieth-century modernism: geometric private gardens, "mixed" parks that combined both natural areas and formal gardens, and the continuity of the garden show as a medium of representation and communication. What was new – besides the functional considerations – was the political and ideological messages that these gardens were supposed to render legible, unlike those of Imperial Germany and the Weimar Republic. One widespread typology of the previous regimes was the "memorial garden", which in many cases served as a venue for mass rituals; by referencing momentous events in the nation's history and its myths of origin, these politically commemorative gardens played an important part in generating a sense of national identity.

Herta Hammerbacher was one of the first women to become a professional garden designer. Her projects from the 1930s onwards, undertaken in collaboration

The Cold War saw gardening
become an ideological battlefield
between East and West.

with the landscape architect Hermann Mattern and the gardener, nurseryman, and writer Karl Foerster,[1] were very successful. Hammerbacher designed gardens for the Reichsgartenschau (Imperial Horticultural Show) in Dresden, Essen, and Stuttgart (1936, 1938, 1939), and after 1945 she was the first woman professor in the faculty of architecture at Technische Universität Berlin. Her style tells of her close affinity with the landscape, a formative factor in the gardens of the post-war reconstruction period and the Economic Miracle of the 1950s and 1960s.

While there were calls for a Marxist-Leninist approach to gardens in East Germany, the number and size of those actually created remained comparatively modest. The green spaces of Stalinallee (today Karl-Marx-Allee), East Germany's signature boulevard built as part of the reconstruction of East Berlin after World War II, took the form of flat parterres laid out before the building facades, with sculptures and architectural ornamentation by way of embellishment.[2] The large public squares in East Germany, such as Altmarkt in Dresden, Alexanderplatz or rather Marx-Engels-Forum in East Berlin, Augustusplatz in Leipzig, and Marktplatz in Neubrandenburg show no signs of green ambition on the part of their designers, and in this respect they are not unlike many West German reconstruction projects.

The politicization of garden design and landscaping in the 1930s and 1940s was followed by a period of optimism and belief in progress and then by the 1970s' emergence of environmentalism, which cast doubt on the model of endless growth. While the Cold War saw gardening become an ideological battlefield between East and West, the dawn of the post-industrial era made the "limits of growth" all too glaringly obvious. Decolonization also made its mark on the garden politics in the age of environmentalism. So can we even talk of any one garden politics? Garden shows such as West Germany's Bundesgartenschau (BUGA, launched in 1951) and East Germany's Internationale Gartenbauausstellung, of which the first edition was

held in Erfurt in 1958, reflected the prevailing political, economic, and aesthetic concerns of each ideological system: while BUGA emphasized model gardens and regional development projects, the East German events set out to showcase the achievements of socialist garden design and landscaping.

As conservative in form and format as German garden politics of the 1950s and 1960s was in both the east and the west, elsewhere in the world, and in the US particularly, the concept of landscape "counter-sites" was gaining ground. Its embedment in the landscape was crucial to its interpretation. Works of land art – which count as landscaped spaces or as gardens up to a point – were an attempt to liberate art from the confines of the studio, museum, gallery, and the academy and express an individualistic concept of freedom. Like (Californian) counterculture of the 1960s and 1970s, the protagonists of land art sought out remote and inhospitable locations in the belief that spatial distance would emphasize their social critique, even if that meant cutting ties to architecture and, above all, urban planning. It was in land art that modernist garden politics found what was perhaps its most radical manifestation. America's urban planning reforms of the same period ventured down a similar path. Among them was Sea Ranch, an alternative community built in 1963 on the precipitous Pacific Coast to the north of San Francisco. Using local materials and traditional building methods, the Sea Ranch project headed by landscape architect Lawrence Halprin created a model housing development that besides being responsive to its site was locally planned and executed, and in that respect posed a genuine (political) alternative to the then social status quo. The horticultural interventions at Sea Ranch were kept to a minimum and took their cues from the existing vegetation (p. 65).

Alongside these escapist residential utopias, there were other projects that set out to forge even stronger links between gardens, cities, and their inhabitants. One of the most noteworthy was Habitat 67, the

→ Renée Gailhoustet and Jean Renaudie, Ivry-sur-Seine, social housing project, Paris, from 1971

↑ Sea Ranch, Sonoma
County, California,
built in 1963, photo:
Iwan Baan, 2015

→ Landscape Park
Duisburg-Nord, former
Thyssen steelworks,
Ruhrregion, Germany

As a political platform for forging a national identity and a sense of community, gardens were extremely important to the processes of decolonization in the post-war period, especially in the southern hemisphere.

utopian housing project financed by the Canadian government, designed by Moshe Safdie, and presented at Expo 67 in Montreal. In his later book *For Everyone a Garden* (1974), Safdie elaborated on his idea of communal living connected to nature in big cities. Unlike Migge's plans for Frankfurt, Safdie envisaged gardens not as a source of food but as social spaces and places in which to savour the world of nature. And unlike Sea Ranch his projects were conceived more as works of architecture than as gardens. During the late 1960s' rapid development and modernization, it was hoped that the insertion of gardens into deprived, densely populated areas with very few green spaces would bring about social improvement. Among those to do some important work in this field was the French architect Renée Gailhoustet, whose magnum opus, the mixed-use city centre of Ivry-sur-Seine, Paris (from 1971), features tightly clustered residential blocks with their own terraced green spaces.

As a political platform for forging a national identity and a sense of community, gardens were extremely important to the processes of decolonization in the post-war period, especially in the southern hemisphere. Selecting examples from such a vast array of regional differences is difficult, but one figure who definitely warrants mention is the Brazilian Roberto Burle Marx, whose public gardens for the new capital of Brasília (from 1956) and interventions in the public space in Rio de Janeiro (Copacabana Beach, 1970) and São Paulo (Parque do Ibirapuera, from 1954) synthesized historical and contemporary influences to produce an unmistakable garden language. Like other landscape architects of the south, Burle Marx can be said to exemplify the way garden politics has broadened, diversified, and globalized since 1945.

Today, the canonical gardens of history play almost no role in providing orientation for new projects. Garden history is regarded as outdated, along with the garden politics of the post-war decades. The politics of the garden in the German context now relates primarily to the transformation of formerly industrial areas, in both the east and the west, as well as the remodelling of industrial wastelands to create parks or extensive recreational areas. Among the examples of this innovation is Internationale Bauausstellung Emscher Park in the Ruhr region (1989–99) and the recultivation of disused opencast mines in the lignite mining districts of central Germany after 1990; these projects are, however, more an outcome of this region's development policies than a result of city planning.

This second decade of the twenty-first century has been a time of great crises and far-reaching socio-economic changes, which has seen gardens and parks gain considerable importance, especially in cities whose parks are well used for recreation, (non-commercial) leisure, and regeneration. This recalls other times in the modern history of gardens – gardening becomes a weathervane for political developments and a gauge on society's most pressing needs. Yet, the extent to which today's gardens can be viewed as political spaces in the sense outlined here, that is to say as spaces that not only represent but actually advance social and political issues, is a question that only the future can answer ●

1   Karl Foerster had considerable influence on later generations of garden designers, including Mien Ruys, who for many years was his apprentice, see p. 114.

2   One exception is the park designed by Helmut Kruse around the high-rise block at Weberwiese, built in 1951–52 to plans by Hermann Henselmann.

Acres of Opportunity: How we became Guerilla Gardeners

When parts of densely built-up New York City fell into decay in the 1960s, few
public green spaces remained, especially for less affluent citizens. In 1973, the artist
LIZ CHRISTY, who lived on the Lower East Side, founded the Green Guerillas,
a group of gardening activists that is said to have invented the "seed bomb".
In this article, co-written with fellow Green Guerillas member DONALD LOGGINS
and first published in 1986, Christy charts how urban development created
a political crisis and catalysed the formation of their pioneering group.

The history of New York City's community gardens shares its roots with the early days of the city. The first European settlers brought with them the tradition of the commons. These areas of land were dedicated to public use and served simultaneously for recreation, as pastures, and for military use, among others. From the beginning of the nineteenth century, however, a large part of these open spaces fell victim to the construction boom in the growing city. Social reformers were concerned that despite the steady rise in population, little open space was set aside for public use.

In the late nineteenth century, the commons was recalled in direct response to the dire needs of the burgeoning lower class. The allotment movement was born when pieces of land were given to the poor via schools, orphanages, or created in public parks. Gardens were seen to ensure food security, fresh air exercise, and education, including the benefits of moral and social training to men, women, and children. Even factory owners provided allotments on their compounds to keep their workers healthy.

During World War I, it became patriotic to grow vegetables. Those left on the home front could do their part with hoes and rows of vegetables in the so-called Victory Garden allotments. These were recalled to meet the needs of the poor with Relief Gardens during the Depression, and glorified again as Victory Gardens with the outbreak of World War II. However, few of these gardens were permanently set aside.

In 1932, only 7.28 per cent of the city was dedicated to recreation; this was less than any of the other ten largest cities in the United States at that time. In addition, there was only one playground for 14,000 children under twelve. In response to this situation, Robert Moses, Commissioner of the New York City Park Department, created 255 new parks over the 1930s.

However, most of these were built in middle- and upper-class neighbourhoods and left out over 200,000 minority youths. Moses also commanded the acquisition of 60 sites – some of them pocket-sized – for revitalization. These included playgrounds that were characterized by a sea of asphalt punctuated by a few shade trees, benches, and iron fences. Maintenance costs were disastrously expensive.

The development of small parks dropped considerably after World War II due to the desperate need for new housing and construction for business and industry. Most of the newly built parks were for the small percentage of those with money to purchase an automobile. Expansive and scenic parkways were planned and designed for this population. And when the city became unbearable, many of these people from the middle and upper classes moved to the suburbs permanently.

During the late 1960s the play lots programme under Mayor Lindsay demonstrated that playground equipment and asphalt could not be dropped into a neighbourhood as if by helicopter and enjoy more than a temporary heyday with the residents. At the time, studies of urban riots stressed that it was important to increase facilities for recreation, so the city was generous with materials to keep low-income residents content with their own neighbourhoods. But what was provided was "in perpetuity" equipment (hardware that is as indestructible as possible) to lower maintenance requirements. The responsibility for maintaining the site later fell to the community. The absence of resident input into the design process and the omission of active participation of the user population in the actual construction diminished the benefits of the gift play equipment. Eventually those play lots – devoid of any living thing, a touch of shade,

↑ Playground in vacant lot, Harlem, 1939

↓ New York City's gridiron street plan, the legacy of an 1807 commission with little designated public space; map published in 1811

→ Two boys playing in a vacant lot strewn with debris, West 91st Street, New York, 1962

In the late 1960s, the number of vacant lots in the city had begun to increase at a phenomenal rate.

colour, or seasonal interest, and with no neighbourhood imprint – disintegrated and fell into the 1970s' waste pile. The side street and rectilinear play lots remained locked into the urban grid, with more hard surfaces such as asphalt, concrete, and temperature sensitive metals conducive to hard play and with little to soften the edges.

In the late 1960s, the number of vacant lots in the city had begun to increase at a phenomenal rate. The combination of arson and landlord abandonment led to rows of burnt-out, bricked up buildings and ugly, refuse-filled vacant lots. As the city's financial condition worsened, urban renewal and other building plans were postponed; an estimated 15,000 acres of the bulldozed areas lay empty. But when we as a society are hurting, we turn to community gardening as a cure.

The Park Association of New York City (later named Parks Council), the city's good government watchdog for urban parks and neighbourhood improvement, began to investigate the potential of vacant land. Their publication of May 1969, *A Little about Lots*, spawned many do-it-yourself, community projects. They also sponsored the Wall Street Flower Show with flowering trees and shrubs, which they later gave to community groups who needed plants from 1970 to 1974. If you were really persistent, you and your neighbours could make "a something from a nothing", and it was alive – better than the play lots built by the city. During the same period Mollie Parnis, the fashion designer, funded the Dress Up Your Neighborhood contest, with cash prizes going directly to the community for their efforts.

In 1973, the Green Guerillas were founded as an organization of multidisciplinary professionals and experienced volunteers. We wanted to improve the quality and quantity of green open space in the city. We also wanted to help community residents with acquiring leases, planning and design, construction and organization. Practical, technical assistance "on-site" was given where it was needed. The now famous seed grenades of old Christmas tree ornaments or water balloons filled with time-release fertilizer and wildflower seeds helped sow a grassroots revolution on "acres of opportunity", as they were dubbed by one Green Guerillas member.

↙ Liz Christy on the site of the first Green Guerillas community garden on Bowery Street at Houston Street, Manhattan, 1973, photo: Donald Loggins

↓ Liz Christy on the same site, 1976, photo: Donald Loggins

→ Instructions for the assembly of a seed bomb, printed and distributed by the Green Guerillas, 1975

People have historically joined together
to grow plants for survival and for pleasure.

Seed Grenade Recipes
(For vacant lot bombing, or how to hide illegal dumping space)

PREASSEMBLE THE FOLLOWING INGREDIENTS:

    A. <u>Old Christmas ball ornaments</u> with metal hangers removed
        Small funnel
        Pelletized, time-release fertilizer
        Small bits of moist peatmoss
        Seeds, suitable for time of bombing and desired effect --list below
        Piece of kleenex or other tissue paper
    B. <u>Small balloons</u>
        Funnel
        Pelletized time release fertilizer
        Water and watering can
        Seeds,see list below

<u>ADD</u>:

    seed and fertilizer to grenade membrane.through funnel.
    In variety <u>A</u>, add wet bits of peatmoss and stuff
             opening at top with small piece of
             tissue paper.
    With variety B, be sure to add the seed and fertilizer
           before adding water.
    Both varieties A and B should be handled with care.

Wet peatmoss
Pellet fertilizer
Seeds
sectional view

<u>INSTRUCTIONS FOR USE</u>:

    Choose a lot that has a fence and is legally inaccessible. Calculate in advance
    how many grenades will be needed to cover the area. Check carefully before
    throwing seed grenade. Observe all normal safety precautions. Perfected throwing
    techniques are: for Christmas ornaments--use underhand throw and for the heavier
    water balloons--an overhand toss.

<u>SEED LIST</u>

for early fall	for early spring	for late spring	for early summer
Soybeans	batchelor buttons	cosmos	sunflower
Clover	dianthus	portulaca	ornamental grass mix
Winter Rye	wildflower mix	zinnia	marigold
cleome	plain old grass	nicotiana	zinnia

P.O. BOX 673, CANAL ST. STA., NYC 10013

The early 1970s produced a handful of really beautiful community open spaces such as Ruppert Green, Village Green, and Asphalt Green, which were privately funded with foundation support and professionally designed. The post-flower power population was growing up. City residents looked more realistically upon the growing decay in housing, urban renewal promises, and the fact that the Parks Department could not take care of our parks.

The first Earth Day in 1970 spawned action on a larger scale, aimed at closing the gap in environmental education. Environmental Action Coalition dealt with street tree and waste issues, while the Open Space Greening Program was initiated by the newly founded Council on the Environment of New York City (CENYC), who provided a tool and book lending library, on-site assistance to groups in all five boroughs, and the publication of practical, how-to information regarding leasing, site evaluation, participatory design principles, and composting that supported groups to recycle organic waste for conditioning the alkaline rubble and making it into a better soil. Exhibitions and lectures were added to provide services to community projects anxious to transform a vacant lot into a tiny Eden. Through CENYC seedlings were distributed, when available. The Green Guerillas and other groups also distributed free trees, shrubs, and other plants from the mid-1970s on.

The urban resident of today has an opportunity to grow vegetables as our ancestors did only in times of depression, war, or crisis. There are over 750 community sponsored open-space project sites in New York City. Some of these are allotments for individual vegetable or flower plots; some are parks involving community participation in maintenance and management; others included recreation. Almost every project was different, but all were embedded in the community.

People have historically joined together to grow plants for survival and for pleasure. The term "community garden" expresses dependency on each other, linked together by basic needs, and giving benefit to each participant. Thus, the cultivation of plants has always been a communal activity, but so far it was only in times of crisis that we turned to community gardening as a panacea ●

Gardening in Times of Crises

by Jochen Eisenbrand

Throughout the twentieth century and up to the present day, gardens have played a central role in times of crisis, war, and displacement – as a means of survival through self-sufficiency, but also a refuge, a bucolic antithesis to the horrors experienced, or as a symbol of resistance.

During the two world wars, gardening was declared a patriotic and thus political act in countries such as the United States, the United Kingdom, Canada, or Switzerland, and accordingly encouraged by state and government institutions.

In war zones and refugee camps, gardens reflect the hope of those affected to eventually be able to put down roots again. The rhythms inherent to gardens, which follow the cycles of nature, are both reminiscent of a time before and look towards a future after the crisis.

Towards the end of World War I, multimillionaire Charles Lathrop Pack founded the National War Garden Commission (NWGC) in the US to foster an interest in gardening among the American public. Posters released in various languages, advertisements in magazines, and educational programmes were targeted primarily at women and children. The campaign had several goals: increase food production, strengthen a sense of community and national spirit, and, finally, to return an increasingly urban and suburban society to its agricultural roots.

James Montgomery Flagg's poster depicts Columbia – a common personification of the US used well into the 1920s – scattering seeds with her hands in a gesture of generosity and with her head held high. The poster's title stylized the prospect of winning the war as a kind of harvest.

In a propagandist effort, the gardens that sprang up on the initiative of the NWGC on private properties, around schools, places of employment, as well as prominent public sites such as the National Mall in Washington, D.C. were called Victory Gardens. In fact, the number of private gardens in the US increased from 3.5 to 5.3 million between 1917 and 1918.[1]

→ James Montgomery Flagg, *Will you have a part in Victory? Every Garden a Munition Plant,* poster for the National War Garden Commission's campaign, USA, c. 1918

In addition to the National War Garden Commission, there were other initiatives to promote horticulture in the US during World War I. While the Victory Garden programme was intended primarily to encourage women to garden, United States School Garden Army (USSGA), funded by the War Department, was aimed at city youths.

In this poster by Maginel Wright Enright, the children of the School Garden Army follow Uncle Sam as he plays the flute. The reference to the legend of the Pied Piper of Hamelin, who lures children out of town, is obvious. Enright worked as an illustrator for magazines and children's books. She was the sister of architect Frank Lloyd Wright.[2]

← Maginel Wright Enright, *Follow the Pied Piper. Join the United States School Garden Army*, 1919

During World War II, planting a Victory Garden was once again praised and encouraged as a patriotic act, not only by the government but also by many employers and educational institutions. According to US Department of Agriculture estimates, towards the end of the war the nation's private gardens yielded around 8 million tons of vegetables – about 40 per cent of the nation's total demand.

→ Children's school Victory Garden on First Avenue between 35th and 36th Street, New York City, c. 1944

In light of the extreme threat posed by neighbouring Germany during World War II, food safety in Switzerland was at risk. Over the previous century, Switzerland – once dominated by farming – had become increasingly dependent on food imports. An *Anbauschlacht* (cultivation battle) was proclaimed in 1940 with the aim to return to complete food self-sufficiency. The first step was to triple the country's arable land.

Noël Fontanet's poster was part of a campaign launched in 1941 by Geneva's Aide volontaire à l'agriculture, appealing to francophone Swiss youths to participate in the cultivation battle. The young, armed with gardening tools, were to see themselves as soldiers on the home front, as their shadows clearly show. Fontanet worked for Allgemeine Plakatgesellschaft APG from 1920 to 1953, a poster company founded in Geneva. He was politically involved in the radical right-wing movement Union nationale.[3]

← Noël Fontanet, *Young People …!! Your Relief! Sign up for the Battle of the Fields*, c. 1942

Two years after the campaign was launched, the so-called cultivation battle in Switzerland produced its first results. Documentary photographs from 1942 show men harvesting potatoes on the square in front of the Opera House in Zurich or at the Federal Palace in Bern. Grains were planted on the Sechseläuten festival green in the middle of Zurich. However, the planting campaigns mainly held symbolic significance; they only managed to reduce Switzerland's dependence on food importation to a limited extent.

↓ "Cultivation battle" in Switzerland during World War II, Zurich, 1942

Photo-reporter Harry Shepherd's July 1942 photograph was published and circulated amongst the British War Office. The image of a garden set in a bomb crater became a symbol of Londoners' resilience during the German bombing campaigns in World War II. The bomb was probably intended for Westminster Cathedral, but had instead fallen on an open square next to it. There, the cathedral's caretaker took the opportunity to plant a vegetable garden in the now exposed soil, which he lined with bricks from the rubble of bombed houses. The official caption reads, "Where the Nazis sowed death, a Londoner and his wife have sown life-giving vegetables."[4]

↑ Victory Garden in a bomb crater, London, 1943

In England, food imports were severely affected during World War II due to German submarine attacks. Here, too, the government launched a national campaign for self-sufficiency through the expansion of horticulture. The poster by Abram Games uses clear imagery to show the journey taken by vegetables, from garden to dinner table, and calls for any fallow land to be converted into a kitchen garden. Games worked as a graphic artist for the British War Office in 1941. A year later he was appointed Official War Poster Artist.

→ Abram Games, *Grow Your Own Food*, 1942

·· every available piece of land must be cultivated

GROW YOUR OWN FOOD
supply your own cookhouse

British war correspondent and photojournalist Lalage Snow has spent many years photographing gardens in crisis zones and talking to those who created them. The two photographs shown here are from her book *War Gardens* (2018).

"I was 16 when I first started working here with my father in the palace gardens. They were beautiful. The king brought different flowers and trees from Europe, and when the palace was being built there were elephants and donkeys to move all the heavy trees. The king was a good man and I remember those days happily. It makes me sad to see this palace in ruins, ruins like my country.

"Now I work in the government vegetable garden for the Afghan National Army based here. I built this garden in the courtyard a few months ago just for fun. I feel like the plants have given me energy. I feel younger and more powerful and think that my hair is losing the white bits. I don't think this garden will last long, it will probably be destroyed. But everything you see growing here is from Paradise and I feel like I'm in Paradise when I'm here. I'm a poor man and sometimes there isn't enough to eat but gardening feeds you in a different way. The soldiers always say they like the greenery as it reminds them of peace. Who doesn't like that?"
– Mohammad Kabir, September 2012

↑ Mohammad Kabir in the Darulaman Palace Garden, Kabul, Afghanistan, September 2012, photo: Lalage Snow

"These beans that I harvest, and the dill there is all I have left. I will sell them at market – it's not enough for my whole life but I sell it so I can buy things for my grandchildren, and children like biscuits. What would you do?" – Lubov Ninolaeuna Lizvinova, 2014

↓ Lubov Ninolaeuna Lizvinova, Simonovka, eastern Ukraine, 2014, photo: Lalage Snow

Dutch photographer Henk Wildschut often focuses his lens on the living conditions of refugees and migrants. The two photographs shown here are from his book *Rooted* (2019), in which he documents gardens in refugee camps.

"Despite the scorching desert heat, I was amazed to find little garden plots complete with garden ornaments all over the Choucha refugee camp when I visited it in Tunisia in 2011. [...] Aware of probably being fated to a long stay in the camp, the inhabitants felt a need to distinguish themselves from the monotonous official surroundings. These little gardens were expressions of resistance to bureaucratically imposed victimhood." – Henk Wildschut, July 2011

↑ Choucha refugee camp, Tunisia, July 2011, photo: Henk Wildschut

"The rose originally comes from Asia and has been bred and cultivated for its fragrance and beauty for thousands of years. The rose has great symbolic value and was used in ancient times to adorn temples and palaces."
– Henk Wildschut, May 2018

→ Spray Rose, Terbol, Lebanon, May 2018, photo: Henk Wildschut

1   Garance Franke-Ruta, "When America Was Female", *The Atlantic* (5 March 2013), online: https://www.theatlantic.com/politics/archive/2013/03/when-america-was-female/273672, accessed 16 January 2023.

Rose Hayden-Smith, "'Soldiers of the Soil': Wartime Gardening Programs of World War I", *Pennsylvania Legacies*, vol. 17, no. 1 (2017), pp. 20–25.

Charles Lathrop Pack, *The War Garden Victorious*. Philadelphia: J. B. Lippincott Company, 1919.

2   Paula E. Calvin and Deborah A. Deacon, *American Women Artists in Wartime, 1776–2010*. Jefferson, NC: McFarland, 2011.

3   Hans Jenny, "Versuch einer schweizerischen Agrarautarkie", *Weltwirtschaftliches Archiv*, vol. 58 (January 1944), p. 88.

4   https://www.seismopolite.com/ecologies-of-resilience-westminsters-bomb-crater-garden-and-the-dig-for-victory-campaign, accessed 16 January 2023.

On the
Lawn

Since the seventeenth century one powerful element has dominated garden design: the lawn. Free of agricultural or horticultural purpose, rather it is a spatial expression of socio-economic status and control, over others and nature. KRIS KOZLOWSKI MOORE explores the lawn's geographical spread and ideological connotations.

A lawn should be no more than an inch and a half tall. Other heights may be justifiable, depending on the season, conditions, and use, but it is that particular measurement that defines what is deemed a lawn. A lawn should also be of one type of grass, free of weeds, sharply edged, and uniform in colour. Today, these tenets are ingrained in our social consciousness, particularly in the West, and lawns have, for many, become synonymous with community, landscape, garden, neighbourhood, and home. Lawns simply exist, naturalized parts of our social and topographical fabric, and through their ubiquity have achieved an almost blind acceptance. Yet lawns are always designed, produced, and managed. Their green guise signifies the natural, while their existence is wholly artificial. As a result, lawns articulate the ideologies of those who create and maintain them.

It was not until the seventeenth century that lawns appeared as symbolic spaces, first amongst the aristocratic estates of England and France. Here, lawns became a landscape that demonstrated the means to possess without the obligation to use. The Gardens of Versailles is a paragon of this early symbolism, with each blade of grass in its vast and orchestrated 800 hectares an image of discipline that testifies to the might of the state and mastery over nature.[1]

In the eighteenth century, open green space became an increasingly fashionable way to signify affluence.[2] This was in part due to the rise of English landscape gardening, led by revered masters like the British landscape architect Lancelot "Capability" Brown, probably the most famous protagonist in the English landscape garden style. His geographical compositions were a departure from the dense formality of earlier renaissance and baroque gardens, instead championing a minimalism that accentuated existing landforms,[3] and where sweeping expanses of undulating grass were intrinsically tied to his vision of Enlightenment modernity.

Ontologically, these swathes of grass were identical to the lawns that exist today, yet the sheer scale

↗ Canaletto, *Badminton House*, designed by Lancelot "Capability" Brown, Gloucestershire, 1748

← Étienne Allegrain, *Louis XIV and his Court on a Promenade in the Gardens of Versailles*, c. 1688

of Brown's pared-back earthworks gave his lawns an almost native aura, a monumentality that implied they were an extension of Nature herself.[4] This simulation of an unaffected terrain concealed Brown's manipulation of the landscape and did much to popularize the idea of the lawn as natural.

Inspired by such gardens, George Washington and Thomas Jefferson transplanted this burgeoning landscaping to North America, employing it at their Mount Vernon and Monticello estates respectively. It was only after 1830, with Edwin Budding's invention of the lawnmower in England, that the lawn and its aesthetic became available to other social strata, equipping them

with the means to pursue the lawn's opulent connotations.[5] While the middle class in Europe embraced the lawn with this new-found technology, its broader popularity is perhaps best attested to Frederick Law Olmsted's 1868 design for Riverside, Illinois – one of the first planned suburban communities in America.

Olmsted remodelled the private English garden, prescribing all houses be set back 30 feet from the road and prohibiting walls between properties. In doing so, he established the prototypical front lawn, an uninterrupted river of green whose unity created the illusion that everyone lived together in a single park.[6] In this way, the lawn-bound residents quite literally but

also implicitly entered a moral contract of maintaining order and beauty. The lawn was no longer only tied to the idyllic imaginary or power, but rather to a new social bond: harmonious, middle-class collectivism.[7] After Riverside, lawns were popularized in the United States at a brisk pace. In 1870, Frank J. Scott published his influential *The Art of Beautifying Suburban Home Grounds of Small Extent*, in which he asserted a well-kept lawn was the definitive feature of a beautiful garden.[8] Scott's rhetoric was followed by further developments that fuelled the lawn's growth: first with the invention of the sprinkler in 1871, and then the rise of mail-order catalogues from which lawn seed could be ordered. These economic, social, and technological conditions did not merely facilitate the growing reach of the lawn but produced it, each laden with agenda.

Following World War II, lawns in America as in Europe experienced another boom: this time motivated by escalating automobile use and the creation of the Interstate Highway System, the development of more efficient maintenance equipment (such as the petrol lawnmower), and a surge in suburban development. In Europe, as well as further east from Singapore to

Aotearoa New Zealand, lawns also continued to proliferate as technology improved, but in hindsight the uptake lacked the same zeal and speed with which the US embraced this picture-perfect landscape. An archetypal American suburb that did is Levittown in Long Island, New York, and between 1947 and 1951 Abraham Levitt and his sons, William and Alfred, built over 17,000 standardized homes.

It wasn't, however, just mass-produced homes they assembled, but a mass ideology as well. Levittown was an enclave of anti-communist thought, economically incentivized for returning veterans, and manufactured in line with the demand for Cold War conformity. Lawns were the perfect fit. William Levitt, credited as the driving force behind Levittown, even declared that no one with a house and a lot could be communist as they would have "too much to do".[9] Part of the Levitt ideology was also racial: Levittown was, by covenant, a white suburb, with the first lease agreements stating that only the Caucasian race could buy there.[10] To those unhindered by racial prejudice, Levittown was marketed as the latest affordable opportunity in the American Dream,[11] further encouraged by the

← Frederick Law Olmsted,
Plan of Riverside, a suburb
west of Chicago, Illinois,
1869

↘ Aerial view of the curved
streets in Levittown,
New York, 1949

assurance that properties with lawns appreciated more than those without. Lawns proved to be the ideal ecological companion to the Levitts's industrialized production, rapidly covering up naked earth and keeping pace with swift construction.

Yet, this calculated ideology was eclipsed by its presentation as "the next best thing" with each plot's cookie-cutter lawn as the welcoming facade. Keeping a neat lawn was now an exercise in pride, ability, and self-respect. To protect these ideals and their appeal, the Levitts's distributed lawn-care advice in residential newsletters and made it mandatory for residents to mow their lawns once a week between April and November.[12] As the labour commitments and social expectations surrounding lawns continued to rise, so too did scepticism about the ramifications of such pursuits.

In 1962, Rachel Carson published *Silent Spring*, a seminal environmentalist text of the twentieth century and one that did much to promote anti-lawn sentiment. In her deft polemic, she unveiled the cataclysmic effects of pesticides: habitat destruction, wildlife decline, food-chain toxicity, watershed contamination, and health risks. She frequently cited the lawn as a site sprayed indiscriminately with only the promise of a greener garden to reassure consumers. Discussing

the application of herbicides to kill crabgrass, Carson seemed mystified by their unwavering use: "Instead of treating the basic condition, suburbanites – advised by nurserymen who in turn have been advised by the chemical manufacturers – continue to apply truly astonishing amounts of crabgrass killers to their lawns each year. Marketed under trade names which give no hint of their nature, many of these preparations contain such poisons as mercury, arsenic, and chlordane."[13] Until this point, the lawn had been proclaimed as the virtuous surrounding of the home, but Carson was suggesting something else, wielding evidence which revealed lawns as spaces subject to corporate motives that were indifferent to the consequences for others.

Predictably, chemical companies retaliated with a slew of pesticide propaganda,[14] yet Carson's indictment of domestic ideals marked an important shift in the history of the lawn. The velvet stripes and joyous families in the advertisements seemed more contrived after *Silent Spring*, and lawns were no longer an accepted symbol of care, but, for some, of carelessness.

Carson, amongst others, drew attention to how lawns are, in their essence, ecological anomalies. Lawns do not grow as a result of geology and climate but are imposed upon the landscape irrespective of

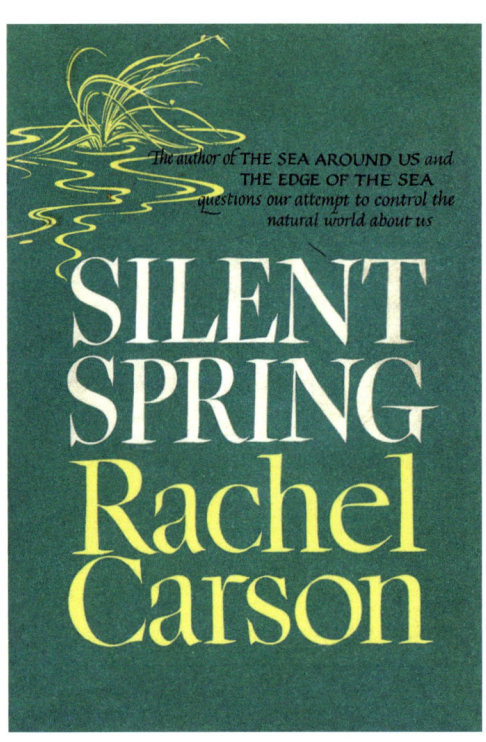

Lawns demonstrate our belief that nature can be moulded at will.

both. And despite their appearance of being static, they are highly active, perpetually ravenous for the concoction of capital, labour, and resources they need to survive. What makes lawns even more perverse is that they must be maintained in reproductive limbo, between adolescence and adulthood – neither allowed to die nor propagate. In this liminal space, growth is stimulated by fertilizers and irrigation, only for that growth to be repeatedly cut down. This promotes the lateral growth critical for a verdant lawn, all the while reducing the ability of the grass to photosynthesize. This increases its reliance on external inputs – another market opportunity.

Today, sustaining such an irrational system requires a bewildering volume of resources. In America alone, there are 40 million acres of cultivated lawn, making it the country's largest irrigated "crop".[15] To water that area, approximately one-third of all residential water is used.[16] Americans deposit 27 million kilograms of pesticides on their lawns annually.[17] A figure amounting to 40 billion dollars are spent each year on lawns.[18] Mowers in America use 800 million gallons of fuel each year.[19] These numbers represent not only the lengths taken to satisfy the rapacious hunger for control but also the colossal matrix of vested interests bound up with the lawn. While America is a drastic example, it is not the numbers that are most concerning. What is perhaps more unsettling, as journalist Michael Pollan writes, is how lawns across the world demonstrate our dogmatic belief that with enough capital and resources nature can be moulded at will.[20] Lawns reveal, in all their green glory, an imperious framing of our planet.

While this incentive of control remains fundamental to lawns wherever they appear – the use of artificial grass in the arid Middle East epitomizes this tenacity – it is not always the case. In Japan, Zoysia grass is one of the most common species used for lawns, although grass lawns are few and far between in comparison to the West. As a warm-season species native to Asia, Zoysia thrives even under fierce sun and is resistant to prolonged droughts. But Zoysia comes with a caveat: it turns brown over winter. For Japan, this modest colour shift merely signifies the changing seasons and is largely embraced.[21] For Westerners, however, it makes the species incompatible with their lust for continuous aesthetic perfection – brown grass is shameful. When Zoysia is used in Western lawns, it is nearly exclusively for its drought-resistant properties; many overseed it with cool season grass in the autumn to ensure a vibrant colour throughout the winter.

← Rachel Carson, First edition hardcover of the book *Silent Spring*, 1962

↙ *Lawn Care*, brochure by O. M. Scott & Sons, Marysville, Ohio, 1975

↘ Children playing on a lawn in front of an apartment building in Potsdam, GDR, 1964

Our need to belong still outweighs our ecological morality, and future buyers of any property with a garden will still look for a well-kept lawn as a signifier of care.

If this seems somewhat bizarre, there is one other activity that, more than any other, epitomizes the efforts still undertaken to uphold the lawn's symbolism: giving it a coat of paint. For most, painting a lawn is an unfathomable notion, an oxymoron melding interior decorating with outdoor space. Yet, many devotees now believe painting their lawn is one of the few infallible guarantees of green perfection in the face of tightening watering restrictions and increasingly frequent droughts. Pollan was right.

Given all that has been said, what do lawns tell us about today? Where do they leave us amidst this cacophony of ideology, environmentalism, and marketing? And why do we stay on the aspirational treadmill of lawns? One possible answer is that it is because lawns remain entwined in a vice of myth, social status, and market hysteria. Our need to belong still outweighs our ecological morality, and future buyers of any property with a garden will still look for a well-kept lawn as a signifier of care. Even here in Europe where lawns are tempered private affairs, lawns sustain the same economic flows and symbolic reflections as their American counterparts. Yet, the treadmill continues because for many the lawn is inextricable from identity, and proponents take something tacit and fundamental about themselves from its potential to exhibit capability and capital.

And if this is what lawns tell us about today, what can we expect of the lawn tomorrow? What is, in other words, the future of the lawn? Truthfully, it's difficult to say. While the lawn's grip remains firm, opponents as early as the end of the nineteenth century have advocated for alternatives to the lawn's hegemony. Arguably the latest of these is the "urban meadow", evident in public parks like Natur Park Südgelände in Berlin and Brockwell Park in London; spaces that at least in part eschew the control so enmeshed in the lawn's existence and instead allow native vegetation to thrive. Collectively, urban meadows demonstrate an important willingness to think outside the confines of the traditional lawn and act as the public display, and in turn the public education, needed if the lawn is to ever be overthrown. Admittedly, however, it's hard to know yet if these wild meadows will be the catalyst many want. Despite a swelling movement behind less contrived green spaces, led by impassioned researchers like Maria Ignatieva and Marcus Hedblom, academics from Perth, Australia, and Uppsala, Sweden, and designers such as James Hitchmough in Sheffield, England, the resistance – at least in the past – has failed to flourish. Indeed, it remains hard to imagine a future without lawns ●

← Natur Park Südgelände in Berlin-Schöneberg

↗ Biodiverse, grass-free tapestry lawn at Ultuna campus, Swedish University of Agricultural Sciences, Uppsala, planted by landscape architect and urban ecologist Maria Ignatieva and research associates, 2016

1 "Gardens of Versailles Had Political and Military Purposes", *Newswise* (31 October 1997), online: https://www.newswise.com/articles/gardens-of-versailles-had-political-and-military-purposes, accessed 1 September 2022.

2 "The Origins of Lawns", *Best Cuts Lawn Care & Landscaping*, online: https://bestcutslawncare.com/the-origins-of-lawns, accessed 1 September 2022.

3 "Why Should We Care about Lancelot 'Capability' Brown?", *Oxford University* (30 August 2016), online: https://medium.com/oxford-university/why-should-we-care-about-lancelot-capability-brown-442e1e724d84, accessed 1 August 2022.

4 "How to Spot a Capability Brown Landscape", *National Trust*, online: https://www.nationaltrust.org.uk/discover/history/gardens-landscapes/how-to-spot-a-capability-brown-landscape, accessed 18 January 2023.

5 "Edwin Beard Budding: Democratizing the Lawn", *personalpedia* (13 July 2009), online: https://personalpedia.wordpress.com/2009/07/13/edwin-beard-budding, accessed 1 August 2022.

6 Whet Moser, "Got a Lawn You Have to Mow? Blame Riverside, Illinois", *Chicago* (23 June 2017), online: https://www.chicagomag.com/city-life/June-2017/Got-a-Lawn-You-Have-to-Mow-Blame-Riverside-Illinois, accessed 12 August 2022.

7 Ibid.

8 Olwen Woodler, "Tired of Pursuing the Perfect Lawn? Consider These Alternatives", *National Wildlife* (1 June 1998), online: https://www.thefreelibrary.com/Tired+of+pursuing+the+perfect+lawn%3F+Consider+these+alternatives-a020925075, accessed 15 September 2022.

9 Richard Lacayo, "Suburban Legend William Levitt", *TIME* (7 December 1998), online: https://content.time.com/time/magazine/article/0,9171,989781,00.html, accessed 16 January 2023.

10 "Developing the American Dream: A Comparison of Levittown, New York and Celebration, Florida", *ARTHIST*, no. 369 (Spring 2018), online: https://documents.pub/document/developing-the-american-dream-a-comparison-of-levittown-itemoryeduatsincludesdocumentsarthistindesign.html?page=1, accessed 16 January 2023.

11 Ibid.

12 Ted Steinberg, "American Green", *Bunk History* (15 March 2006), online: https://www.bunkhistory.org/resources/4634, accessed 10 October 2022.

13 Rachel Carson, *Silent Spring*. New York: Houghton Mifflin, 1962, p. 80.

14 David Ord, "Forty-seven Years since Silent Spring: What Has Changed?", *Ecologist* (11 August 2009), online: https://theecologist.org/2009/aug/11/forty-seven-years-silent-spring-what-has-changed, accessed 16 January 2023.

15 Eric Holthaus, "Lawns Are the No. 1 Irrigated 'Crop' in America. They Need to Die", *Grist* (2 May 2019), online: https://grist.org/article/lawns-are-the-no-1-agricultural-crop-in-america-they-need-to-die, accessed 12 October 2022.

16 "Outdoor Water Use in the United States", *Environmental Protection Agency*, online: https://19january2017snapshot.epa.gov/www3/watersense/pubs/outdoor.html, accessed 4 October 2022.

17 Donald Atwood and Claire Paisley-Jones, "Pesticides Industry Sales and Usage: 2008–2012 Market Estimates", *Environmental Protection Agency* (2017), online: https://www.epa.gov/sites/default/files/2017-01/documents/pesticides-industry-sales-usage-2016_0.pdf, accessed 4 October 2022.

18 "America Spends More Money on Lawn Care Than Foreign Aid: Why We Need Less Lawn", *People Powered Machines* (16 December 2008), online: https://peoplepoweredmachines.wordpress.com/2008/12/16/america-spends-more-money-on-lawn-care-than-foreign-aid-why-we-need-less-lawn/, accessed 16 January 2023.

19 Jiahn Son, "Lawn Maintenance and Climate Change", *Princeton Student Climate Initiative* (12 May 2020), online: https://psci.princeton.edu/tips/2020/5/11/law-maintenance-and-climate-change, accessed 26 September 2022.

20 Michael Pollan, "Why Mow? The Case against Lawns", *New York Times Magazine* (28 May 1989), online: https://michaelpollan.com/articles-archive/why-mow-the-case-against-lawns, accessed 1 August 2022.

21 Anika Ogusu, "Japanese Lawn – In Comparison with Western Lawns", *Real Japanese Gardens* (23 June 2017), online: https://realjapanesegardens.wordpress.com/2017/06/23/japanese-lawn-in-comparison-with-western-lawns, accessed 11 April 2022.

This is an extended version of an article that first appeared in *Real Review* #11, 2021.

Nature's Helpers

by Jochen Eisenbrand

Gardening also means nurturing the desirable and removing the undesirable. Whether it is advertisements for fertilizers, insecticides, or garden tools, the promise is always the same. If you use the right fertilizer, eliminate all pests, work with the best tools, you will, according to the prophecy of the manufacturer, attain one thing above all: tranquility in your garden – an escape from the toil of gardening and the neighbour's critical eye.

Taking a look at the garden products created by a range of different designers, they have primarily tackled gardening with the aim of making work easier and more convenient.

In the mid-nineteenth century, German chemist Justus von Liebig realized that plants grow by extracting mineral nutrients from the soil, and that these nutrients can be added back to the soil through targeted mineral fertilization. He thus created the basis for agricultural chemistry and the fertilizer industry. The most important mineral fertilizer to date – nitrogen, in the form of ammonia-based compounds – was produced in large quantities from 1913 onwards. In this attractively designed brochure from the same year, featuring a carefully manicured, park-like garden, fertilizers are also advertised for private use at home.

↗ *Fertilization in Vegetable and Flower Gardens*, Booklet published by Deutsche Ammoniak-Verkaufs-Vereinigung, 1913

↙ Advertisement by German chemical company BASF for urea-based fertilizer, c. 1922

Aus einem Harnstoff-Düngungsversuch auf dem Limburgerhof bei Mutterstadt, Rheinpfalz.
(Aufgenommen nach der Natur mit Agfa-Platte).

The dustable powder Arbitex Lindane by VEB Berlin-Chemie was used in the GDR in the 1950s to combat the turnip root fly. Like many advertisements for insecticides, this one closely resembles wartime iconography. In fact, there are many overlaps between pest control and chemical warfare, both in research and industry.[1] The gardener's weaponry also often resembles the soldier's weaponry. Before World War I, in Germany, for example, explosives were widely used, and marketed until the 1930s as a labour-saving alternative to digging up and loosening the soil.[2]

→ Advertisement by VEB Berlin-Chemie for Arbitex Dust, 1966

Glück
und Ruh
findest Du
im
gepflegten
Garten

In the 1950s, designer Andreas His was commissioned to redesign the packaging of the herbicides and pesticides sold by Basel-based company Geigy. His minimalist design won a prize in 1957, with the jury particularly praising the "toxic" appearance of his colourful packaging.

↗ Andreas His, Packaging designs for various pesticides produced by Geigy, Basel, 1954–56

↗ Hans Aeschbach, Poster
for Lonza fertilizer, 1960

Wasserarmaturen und Geräte aus thermoplastischen Kunststoffen

 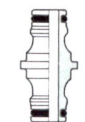

Zwillingskupplung für das Kuppeln von, mit Schlauchstücken ausgerüsteten Wasserschläuchen.

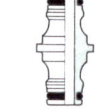

Material:
Polystyrol schlagfest

Drillingskupplung für das gleichzeitige Anschliessen von zwei Geräten.

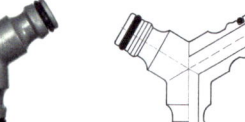

Material:
Polystyrol schlagfest

Gardena, a company founded in the early 1960s, commissioned designers Dieter Raffler and Franco Clivio – who had just completed their design studies at Hochschule für Gestaltung Ulm – to create a novel water hose-connecting system. It went on to win many awards. While Raffler turned to other company commissions from 1983 onwards, Swiss-born Clivio continued to design new products for the Ulm-based firm for decades to come.

← Dieter Raffler and Franco Clivio, Two- and three-way couplings for the original Gardena System, 1967

In 1973, Gardena was the first manufacturer to introduce a 12-volt battery for garden tools, making electric cables obsolete. Since 1990, many of the company's garden tools are operated with the same battery. The wide range of gardening and especially lawn care tools available only reflects the importance of a well-tended garden in post-war society.

↓ Promotional image for Gardena

Wasserarmaturen und Geräte aus thermoplastischen Kunststoffen

Farbgebung

↑ Dieter Raffler and Franco Clivio,
Colour scheme for the original
Gardena System, 1967

We sell leisure...we just call it a lawnmower

mowbot™...the world's only truly automatic lawnmower

Impossible! Not really. ☐ Until you see MOWBOT, you'll never believe it. Turn it on. Electronically-controlled MOWBOT mows and steers itself around your lawn until the grass is completely cut. Leaves flowers, trees, and shrubs untouched. ☐ Safe? Sure. Double rotary blades won't throw grass, stones, or foreign objects. And the MOWBOT mower stops automatically at the touch of any obstruction. ☐ Rechargeable electric power makes MOWBOT a good neighbor. No noise. No fumes. So, if you have nothing to do while MOWBOT mows your lawn, relax; you'll think of something.

MOWBOT INC., 9 Hackett Drive, Tonawanda, New York 14150
If your favorite lawn and garden center or hardware store doesn't yet carry MOWBOT, write for the name of your nearest distributor.

JULY 1969   147

As early as the beginning of the 1970s, the first robotic lawnmower became available on the market. Advertisements for the Mowbot finally declared the garden a work-free zone.

↗ Mowbot Model 900, Mowbot Inc., New York, design 1970; production from 1971

1   Edmund Russell, *War and Nature: Fighting Humans and Insects with Chemicals from World War I to Silent Spring*. Cambridge: Cambridge University Press, 2001.

2   Dresdner Dynamitfabrik (ed.), *Vervollständigtes Handbuch über Anwendung und Vorteile des "Romperit C" Sprengkultur-Verfahrens*. Dresden: Dresdner Dynamitfabrik, 1913.

The Invention of Leisure – From Park Bench to Lawn Lounger

by Nina Steinmüller

The garden city and allotments for growing vegetables were among the concepts to emerge in relation to the appalling living conditions in the rapidly industrializing cities of the West during the late nineteenth and early twentieth century. Whether a public recreation area or a secluded private retreat, the garden came to symbolize relaxation and regeneration.

This transformation can be inferred from changes in the design of garden furniture, as seen in the variations produced for recreational purposes. Among them was the deck chair, whose name recalls its origin as seating on the sun decks of ocean-going liners; the lounger; folding chairs; and furniture that was sturdy but light enough to be moved around.

Both British and Prussian iron foundries were producing large numbers of cast iron furniture even in the late eighteenth century. The mid-nineteenth century then saw the proliferation of lightweight outdoor furniture, especially in France, made of slender, bent iron bars, which as a material was more weatherproof than wood and lighter than cast iron. These chairs were also cheaper and easier to manufacture and could be produced in any number of variations.[1] Although lightweight and readily movable, they were sturdy to sit on. This was thanks to the construction of the chair, in which the seat rests on four U-shaped iron bars whose ends converged at each corner to form the chair legs, removing the need for bracing. So popular were these chairs and benches that numerous competitors in both France and Germany soon began manufacturing similar products of their own.

↗ Garden chair, c. 1850–1900

→ Elegant French chairs in the spa gardens of Vichy, no date

→ Adirondack Chairs in a backyard rose garden, Harrisburg, Pennsylvania, 1948

The most widely used item of garden furniture in the USA is almost certainly the Adirondack Chair. While countless variations of this armchair have surfaced over the years, the basic structure has remained the same. With its deep, backward-sloping seat resting on rear legs angled out, a tall backrest made of wooden slats, and wide armrests, the Adirondack makes for a relaxed sitting position with a low centre of gravity. However, it is not so easy to move as the chair is both heavy and bulky.

The Westport Chair shown here is considered a precursor of the Adirondack, but differs on account of its narrow backrest made of a single wooden board. Its inventor is thought to have been Thomas Lee of Westport, Upstate New York, whose prototype was patented and produced for home use by his friend, Harry C. Bunnell.

↖ Thomas Lee, Westport Chair, c. 1903

"Beautifully shaped, practical, solid, indestructible"[2] – that is how furniture manufacturer MEWA prized the Landi Chair as it rolled off its production lines in the late 1950s. Designed by Hans Coray in 1938, the chair owes its name to the 1939 Swiss National Exposition in Zurich (popularly known as the Landi) where it served as seating. It has remained in production to this day, and since 2014 in a modified version at Vitra.

Aluminium has been used in furniture making only since the 1920s. The Landi Chair, alongside designs by Marcel Breuer, is among the most prominent examples of early European aluminium furniture. The weatherproof material, the shell with punched holes to reduce weight, and its stackability makes it ideal for outdoor use. The 1500 Landi Chairs provided as seating for weary visitors at the Swiss National Exposition were later sold off at 15 Swiss francs a piece at a time when they cost 28 francs new.[3] This explains how so many of them ended up on restaurant patios, balconies, and in private gardens. The minimalist design assured the chair enduring success, which has remained unchanged to this day.

← Hans Coray, Landi Chair, 1938

Cheick Diallo is one of the best known furniture designers in Africa. A trained architect and designer, he lives between Mali and France; his steel and nylon furniture is made exclusively by crafts-people in Bamako. The seat, backrest, and frame of his Ségou rocking chair are all one piece, rather like the Eternit Chair by Willy Guhl. Here, however, the material is not rigid, but woven out of colourful nylon strips strung between two steel frames. While Malian crafts-people of different trades do not normally work together, Diallo's combi-nation of steel and traditional African weaving techniques calls for a collective manufacturing process.

↗ Cheick Diallo, Ségou, 2002

The architect Aino Aalto designed numerous items of indoor and outdoor furniture, most of which were for projects by her architect husband Alvar Aalto. This garden lounger made of painted birch was probably created in connection with Alvar Aalto's design of Villa Mairea for the industrialists Maire and Harry Gullichsen in Noormarkku, Finland. With wheels instead of rear legs and two hand grips on either side of the footrest, it can be moved around like a wheelbarrow. It has sold in various colours and with a choice of additional short armrests.

↖ Aino Aalto's lounger with a padded cover in the garden of Villa Mairea, no date

The Fauteuil 300, developed by plastics engineer Henry Massonnet and manufactured by French company STAMP, was not the first chair to be injection-moulded as a single piece in a single step, but it would certainly pave the way for numerous imitators and define the design of what is now the world's most widely used chair: the ubiquitous Monobloc is a stackable, weatherproof, cheap, and durable plastic chair made of lightweight polypropylene, which is as likely to be found in parks and gardens as it is in restaurants or on the beach. The 1973 oil crisis, however, put a dampener on Massonnet's success, and the 1980s saw other companies learn how to make affordable plastic chairs of their own. STAMP's advertising brochures of the period push the ideal of the garden as a place of rest and relaxation, whereas little work as possible is done.

→ Henry Massonnet, Fauteuil 300/Monobloc, 1972

↘ Brochure of Monobloc manufacturer STAMP, 1980s

The Eternit Chair by Willy Guhl is regarded as an icon of modernist garden design in that it combines a new material for furniture with a contemporary form: a single sheet of still-damp fibre cement is bent into a chair shape, and the two ends are then joined together to form an endless loop. The slightly convex underside and anatomically shaped seat make this a solidly built rocking chair, which, true to the spirit of post-war modernism, enables unconventional ways of sitting and epitomizes the organic language of forms of the 1950s. The manufacturer Eternit used the same durable material to make planters, stools, and troughs, which still feature in many a private garden. The 1990s saw the asbestos previously used in the fibre cement replaced with alternative materials due to the former's carcinogenic properties.

← Image from the Eternit price list, 1967

↓ Willy Guhl, Eternit Chair, 1954

1   Georg Himmelheber, *Möbel aus Eisen: Geschichte, Formen, Techniken*. C.H. Beck, Munich (1996), S. 55 f.

2   Advertisement by MEWA, Wädenswil, 1959–60.

3   See copy of an invoice belonging to Henriette Coray, the designer's widow.

Testing Grounds

How is a garden designed? Those who believe that garden design begins and ends with choosing the right selection of plants for the location have overlooked the fact that gardens, unlike designed objects and interiors, are never actually finished but are rather in a perpetual state of becoming. All gardeners experience failure, enjoy success, and negotiate and collaborate both with other people and with nature. What gardens demand of us is lifelong learning. Here, personal and social fantasies are tested, experiences are processed, insights are gained, and new relationships to the environment and to nature are defined.

Just how varied the approaches to this can be is evident from the eight case studies presented here. As examples of how important contemporary and historical figures have aesthetically staged their own gardens, they each represent different standpoints and intentions that have taken on material form. While one is a place for coming to terms with personal trauma, another is a vehicle of well-being for everyone. While one gardener has dedicated his whole life to nature conservation and the scientific study of plants, another views them as symbols of oppression or empowerment, and another still belongs to the age-old tradition of cultivating plants for medicinal purposes. The inevitable tension between nature and culture, between gardens and architecture is a constant theme of urbanist discourse. But as these contemporary examples in particular show, gardens today are also an important field of experimentation for visions of the future. Instead of serving as miniature worlds or refuges from reality, they have become promising prototypes in which to put socially and environmentally fairer societies or business models to the test.

With her tireless engagement both in the garden and in publishing,
Dutch landscape designer, plantswoman, and publisher
MIEN RUYS (1904–99) set out to democratize garden design.
by Leo den Dulk

Gardens for Everyone

One of the very few garden architects to play a major role in modernism, Mien Ruys (1904–99) developed a functional approach to the meaning of gardens and to gardening in the modern city. Her oeuvre spans more than 70 years, with her most important works created between 1938 and 1978 and the designs she produced in the post-war reconstruction era considered her most innovative.

Although Mien Ruys was raised at her father's internationally renowned Moerheim nursery at Dedemsvaart, the Netherlands, she was not content with being just a plantswoman; she wanted to become a garden architect, "if that is possible for a woman", she wrote in her diary. She started her training in the United Kingdom, but a greater influence was the new German approach to garden design related to the modernist movement in architecture, represented by architects such as Peter Behrens and the Taut brothers who stressed the social importance of gardens and parks. German nurseryman Karl Foerster, co-founder of the magazine *Gartenschönheit* who advocated

for the use of perennials in a much more naturalistic way than the formal carpet bedding of the Jugendstil garden, would visit Moerheim regularly to exchange plants and ideas. On his instigation, in 1929, Mien Ruys took a semester of *Gartenkunst* at the brand new Institut für Gartengestaltung at Berlin-Dahlem. There she became convinced that she had found what she had been looking for: "A movement in garden architecture. [...] A 'new spring', and I am part of it." She then took a semester studying architecture in Delft in 1932, where she "started to discover the larger context of relationships between urbanism, architecture, and garden and landscape architecture".

Mien Ruys developed her approach to the garden through her interests in modernism. She saw the garden as a cultural phenomenon, bringing together that which nature has to offer with the human need for colour, variation, and aesthetics. As she claims by using perennials: "Humans in their daily lives have become so estranged from nature that they seek compensation in a garden with a naturalistic appearance."

→ Mien Ruys in her own gardens at Dedemsvaart, 1950s

← Mien Ruys (landscape architecture) and Gerrit Rietveld (architecture), De Ploeg weaving mill, Bergeijk, built 1956–59, photo from 1965

↙ Mien Ruys, Confection borders in the trial garden of Tuinen Mien Ruys, Dedemsvaart, 2005

↓ Mien Ruys, *Groep Zomerbries*, drawing to illustrate a "standardized or confection border of sturdy flowering perennials for a sunny border on good garden soil or on dry, well fertilized soil", Moerheim nursery summer catalogue, 1958–59

Groep „Zomerbries"

Sterke forse boeket voor zon op goede tuingrond, maar ook op droge – mits goed bemeste – zandgrond.

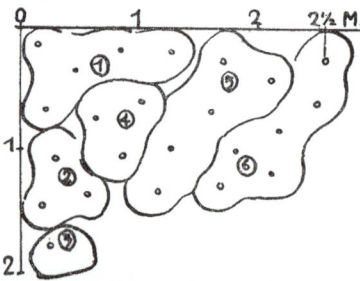

Niet altijd is een bijzondere plant voor de tuin hoofdzaak. Gewone sterke, van ouds bekende planten kunnen maanden lang een dagelijks weerkerende vreugde geven als ze in vorm zowel als kleur een harmonische groep vormen.

Nr	Aantal	Soort	Prijs
1	5	Helenium Moerheim Beauty	*f* 3,25
2	3	Aster dumosus Prof. A. Kippenberg	2,40
3	1	Geranium platypetalum	0,75
4	3	Achillea filipendulina	2,25
5	5	Salvia sup. lubeca	3,75
6	6	Geum borisii	3,90
			Samen *f* 16,30

Prijs voor de gehele groep „Zomerbries" *f* 15,50

Mien Ruys developed her approach to the garden through her interests in modernism. She saw the garden as a cultural phenomenon, bringing together that which nature has to offer with the human need for colour, variation, and aesthetics.

By the early 1930s, Mien Ruys had become a big name in Dutch garden architecture. She published articles in magazines and wrote popular books about gardening. In a 1942 issue of *De 8 en Opbouw*, an avant-garde magazine for modern architecture, she explains her ideas about the role of garden design in the urban environment. Both private and communal gardens should create contrasts with the built environment, she writes. Communal gardens demonstrated the ideal of collectivism cherished by the modernist architects and urbanists she worked with. Together they created communal spaces in which children had a safe playground, where all occupants could enjoy sunlight and fresh air, and the planting schemes reflected the changing seasons.

In the winter term of 1942–43, she started teaching Garden Art Practice at the Amsterdam Academy of Architecture; she was the first female professor to teach at the academy. She also taught classes at Delft University of Technology's Faculty of Architecture and Wageningen Agricultural University. Though it was with her *Het vaste planten boek* (The Book of Perennials, 1950) that she popularized gardening. After Ernst Graf Silva Tarouca's *Unsere Freiland-Stauden* (Our Outdoor Perennials, 1910), hers was an up-to-date handbook for the use of perennials, with a new assortment tried and tested in her own gardens at Dedemsvaart. With the publication of her quarterly journal *Onze Eigen Tuin* from 1955, her influence grew even larger.

After World War II Mien Ruys created a new type of rational, standardized border design for application in smaller private gardens. But while she continued to design classical modern gardens for mansions and villas, she increasingly focused on public green space, including communal gardens in new garden cities and gardens for factories, hospitals, and schools; for her communality was the most important aspect of garden design. As a member of the modernist architecture movement De 8, she participated in the design of Nagele village, situated on newly reclaimed land in the Dutch Ijsselmeer. Among the icons of post-war urbanism, Nagele village was realized with the collective involvement of virtually all the modernist

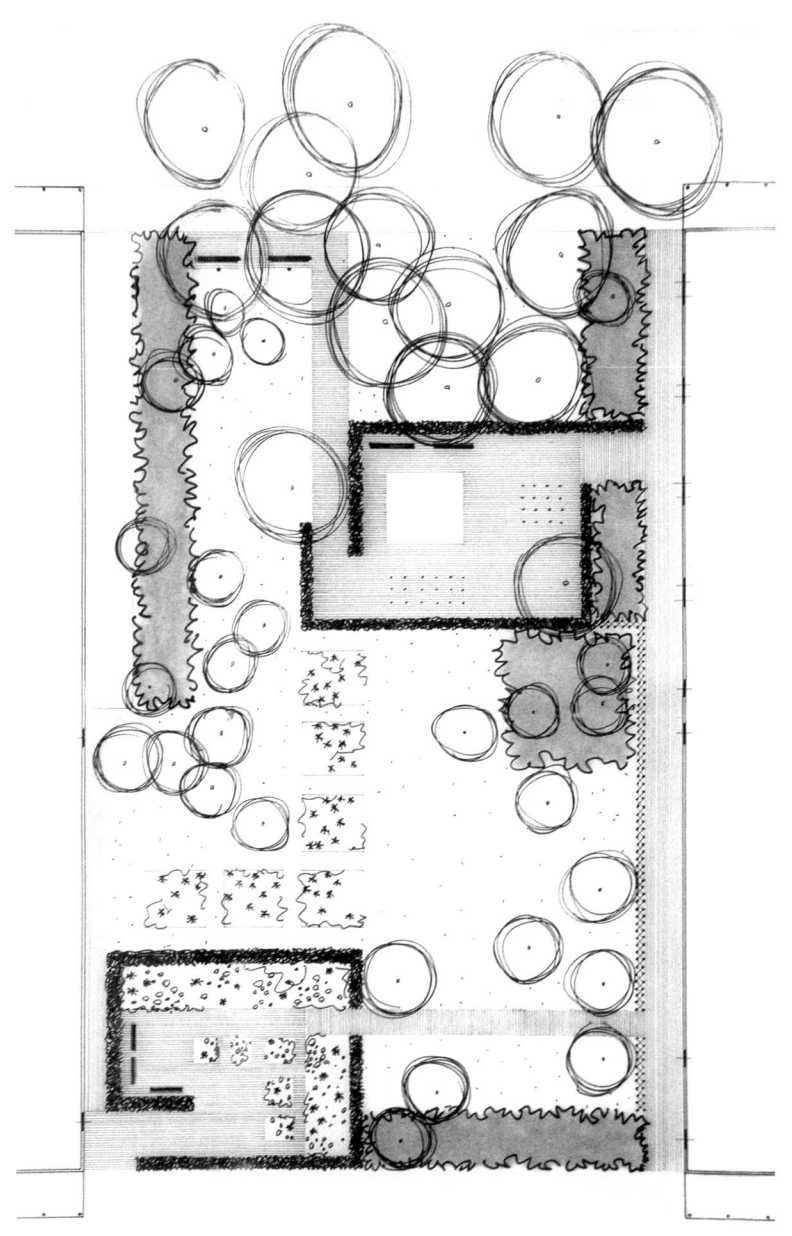

↑ Mien Ruys, Final design of the communal garden at Dikninge, Amsterdam-Buitenveldert, 1962, ink on tracing paper

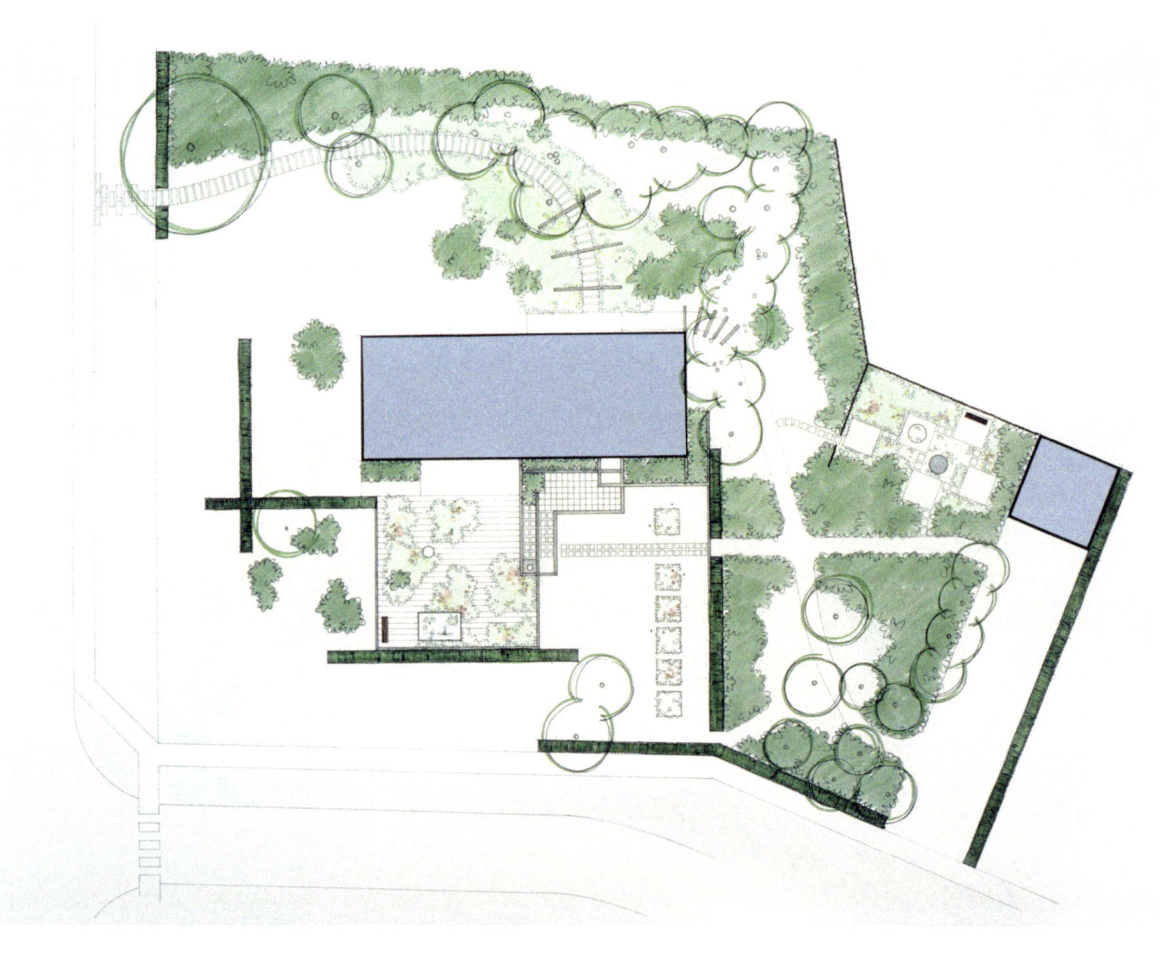

← Mien Ruys, Garden design for the house-of-fice of modernist architect Abe Bonnema at Hardegarijp, 1969, ink and felt-tip pen on paper

↙ Mien Ruys, Abe Bonnema's garden, 1969

Here, materials typical of the 1960s were used: railway sleepers as steps and for a platform, washed-gravel tiles for a garden path and a terrace.

↓ "Standard borders" with tried and tested perennials, Tuinen Mien Ruys, Dedemsvaart, 1970s

architects and urbanists of the day, from Gerrit Rietveld to Jaap Bakema, who finally felt able to realize their ideas without too much outside interference. Other major projects in the same vein included the Western Garden Cities and other parts of the Amsterdam Expansion Plan. What all these projects had in common was a well-coordinated spatial structure, plenty of green space, and a willingness to embrace innovations in design for housing and communal facilities, such as for shops, churches, and schools.

From the early 1960s on, architecture and urbanism started to change. More opportunities for individual expression, pluralism, and small-scale projects were demanded. As Mien Ruys had always designed her plantings on the human scale and with the local situation as a starting point, it wasn't hard for her to adapt to these new principles of structuralism. In other design disciplines, these demands were translated as interconnected geometric shapes: circles, squares, rectangles, hexagons, octagons, and cell structures. Mien Ruys introduced round and square terraces through her planting spaces, rose plots, and hedges.

Although stepping back as designer-in-chief in 1970, she still retained a strong influence in Buro Mien Ruys. And her experimental gardens remain as showcases of her garden architecture, including Tuinen Mien Ruys at Dedemsvaart, where the history of her designs, plant trials, and the garden designs by others are on display. The older parts of the garden are also among the first to be protected under the National Heritage Law.

Mien Ruys's work might best be characterized as a synthesis of German and British influences. The function of the garden, park, and landscape is always related to human activities and needs and her treatment of the space-mass relationship reflects this, never theatrical, dramatic, nor overwhelming, but always in balance with architecture, its environment, and the land. This is not to imply that she was afraid of broad gestures, rather that they were always a means and never an end in itself. A central element of her designs are rich planting schemes with an abundance of colour and shape, set against clear, sober lines of hard landscaping and the modern architecture of the building. Her surviving work still shows this remarkable craftsmanship.

With an oeuvre of over 1000 documented designs – some published internationally as early as 1953 – she has been a major influence on professional and amateur, local and international gardeners and landscape designers alike, such as Piet Oudolf, Jacqueline van der Kloet, and many others, who will readily admit they are greatly indebted to her work ●

↑ Mien Ruys, Working seated in her garden at Dedemsvaart, 1950s

↖ Mien Ruys and Arend Jan van der Horst, Garden design for a modernist private home (architect, Hein Salomonson), Amsterdam, built 1970, photo from 2016

The landscape architect ROBERTO BURLE MARX (1909–94) not only
modernized Brazilian garden design, but made a substantial contribution
to the protection of the rainforest with his research into native flora.
by Nina Steinmüller

The Collaborator

Roberto Burle Marx was one of the great landscape architects of the twentieth century. He was a man of many facets who studied painting and singing but was also active as a poet, sculptor, and graphic artist. Starting in the 1930s, he revolutionized Brazilian garden design, opening it up to the manifold influences of the European avant-garde. His more than 2000 landscaping projects took him to Brazil, Argentina, Chile, and other South American countries as well as to Europe and the United States. Alongside his work in private gardens, Burle Marx's primary interest lay in the design of urban space, which – in collaboration with major Brazilian architects such as Lució Costa, Oscar Niemeyer, and Affonso Reidy – he sought to transform with public plazas and promenades. Less well known, but no less momentous, was his active dedication to the protection of the environment and its biodiversity.

When Burle Marx's German father took the whole family to Berlin for a year and a half in 1928–29, he unwittingly gifted his then 18-year-old son an experience that the latter would later describe as an "awakening". For it was in the Botanical Gardens of Dahlem that the young Burle Marx first encountered plants from the Amazon rainforest, plants that were virtually unknown at the time in Brazil. Realizing that the gardens of his home country were all but devoid of native species, the budding landscape architect made it his mission to remedy this situation, in doing so becoming a passionate defender of biodiversity. He and his team would later undertake numerous expeditions into the Amazon basin to collect and study rainforest flora.

Like architecture and design, landscaping entails collaboration, and Burle Marx built up a large network of creative partners, including several architects,

→ Gardens and lakes at the entrance of the Sítio Roberto Burle Marx in Barra de Guaratiba, Rio de Janeiro

botanists, gardeners, and artists who would play a key role in the realization of his artistic and horticultural visions. Among his most trusted assistants were José Tabacow and Haruyoshi Ono, who accompanied him on his field trips into remote parts of the rainforest and kept records of their journeys. Ono worked with Burle Marx for over 30 years, taking over as director of Studio Burle Marx & Cia. Ltda. upon Burle Marx's death. In 2017, when Haruyoshi passed away, his daughter Isabela Ono became the Executive Director of the studio. Two years later, she founded – together with a board – the Burle Marx Institute, a non-profit organization with the mission of preserving the vast archive of landscape design drawings, writings, project records, and photographs, and to promote and disseminate the legacy of Roberto Burle Marx and his collaborators. The documents tell the story of Burle Marx and his team and their critical engagement with "the history and cultural heritage of Brazil, with the modelling of urban space and how it is used and accessed, with civil rights, public policy, and socio-environmental transformation, and with development and the impacts of development on the quality of life for the local population".

According to Isabela Ono, the importance of Burle Marx's expeditions to investigate the tropical flora of Brazil cannot be overstated: "Throughout his life, Burle Marx worked to identify and cultivate those tropical plant species that had not yet been properly studied in Brazil, more than 50 of which were

↖ Roberto Burle Marx, *Jardim com pandanus* (Garden with Pandanus), 1934/37, india ink on paper, 64 × 79.5 cm

↑ Roberto Burle Marx and Haruyoshi Ono at the Burle Marx Studio, Rio de Janeiro, early 1990s; © Burle Marx Institute archive

↖ Research trip to Araçuaí and Januária, in the north of Minas Gerais, Brazil, 1970s

→ Roberto Burle Marx, Plan for the rooftop garden of the Ministry of Education and Health, Rio de Janeiro, 1938, Gouache, 105.5 × 52 cm; © Burle Marx Institute archive

Gardens are organized
nature, in which the artist's aim
is to highlight the beauty of
colours and forms, rhythm and
beautifully arranged masses.
It means establishing harmonies
and creating contrasts to make
up a whole mesh of elements,
all of them indispensable.[1]

discovered by him and his collaborators. His groupings
and arrangements of native plants in private and
public gardens and green spaces introduced the people
of Brazil to their country's native vegetation, enabling
them to learn more about its biomes. His reflections
on a holistic view of nature were central to his pioneering struggle to protect the environment."

A fierce and tireless critic of forest clearance for
logging and road-building, Burle Marx warned of
the dire consequence of this practice for an intact ecosystem, but: "Even more powerful than his public
pronouncements were his own environmentally motivated actions, specifically those landscape projects
that made for greener and healthier cities and
gave ordinary city-dwellers an aesthetic experience
to enrich their everyday lives."

In 1949, Burle Marx and his brother purchased the
Sítio Santo Antônio da Bica, a farm in Barra de
Guaratiba on the southern fringes of Rio de Janeiro,
where he would henceforth live. The Sítio Roberto
Burle Marx, which became his test-bed for artistic and
botanical research as well as a private refuge to paint
and receive friends, is now a UNESCO World Heritage
site. Among the more than 3500 plants that Burle Marx
and his assistants collected and cultivated in their
various greenhouses were many that he himself had
discovered. The garden of the Sítio became an experimental landscape laboratory in which the impact
of volumes, colours, and shapes could be tested and
the play of textures, contrasts, and perspectives
probed. Here, Burle Marx could study the development
and care of the plants that he had collected prior
to incorporating them into his own garden designs.

One of Burle Marx's earliest surviving projects
is the rooftop garden of the Ministry of Education
and Health in Rio de Janeiro, the design for which was
commissioned in 1938. It is a shining example of the
composure of his gardens, how well they harmonized
with the architecture of the period – in this case,
a building designed by a team of young architects that

A garden is a complex of aesthetic and sculptural intentions;
and the plant is, to a landscape artist, not only a plant – rare,
unusual, ordinary, or doomed to disappearance – but it
is also a colour, a shape, a volume, or an arabesque in itself.[2]

← Roberto Burle Marx
painting, presumably
the plan for the garden
in the grounds of the
US embassy in Rio de
Janeiro, no date

↗ Roberto Burle Marx,
Rooftop garden of the
Gustavo Capanema
Building, the Ministry of
Education and Health
headquarters, photo:
Marcel Gautherot,
c. 1955

→ Roberto Burle Marx,
Pavement design
with irregular forms,
Avenida Atlântica
at Copacabana
beach, Rio de Janeiro,
1970; © Burle Marx
Institute archive, photo:
Haruyoshi Ono

included Oscar Niemeyer, headed by Lúcio Costa, and with Le Corbusier acting as project consultant. Burle Marx continued the greening of the forecourt with organically shaped beds on the rooftop garden, where he orchestrated an array of plants of varying heights in elegant curvaceous beds. His draft designs for projects such as this generally took the form of two-dimensional gouaches that clearly show the influence of Henri Matisse and Joan Miró.

Improving inner-city living conditions was a matter especially close to Burle Marx's heart. The parks that he designed for Rio de Janeiro, São Paulo, and elsewhere gave Brazilians a chance to experience their native flora up close, while the abstract forms, also seen in his paintings, were proposed for large areas of paving, creating true artistic panels on urban surfaces. Today's ostensibly "new" approach to improving the climate through urban planning projects was anticipated long ago by Burle Marx's work of the 1970s, as Isabela Ono explains: "The creative process underlying his landscape projects and the choices that he made can be observed all the way through his work, from the first drawings to the final design. The historical projects possess both an artistic and a humanistic dimension in that they create urban spaces for shared recreation and relaxation, spaces that connect cityscapes and landscapes, that strengthen the bond between inhabitants while remaining sensitive to their individual and collective experiences. These themes are still present in the contemporary 'liveable cities' discourse and have become especially pertinent in today's global context." ●

Unless otherwise stated, all the quotations are of Isabela Ono, Burle Marx Institute, Rio de Janeiro.

1   Roberto Burle Marx, quoted from *Landscape Film: Roberto Burle Marx*, director: João Vargas Penna, Kino Lorber, 2019.

2   Roberto Burle Marx, "A Garden Style in Brazil to Meet Contemporary Needs: With Emphasis on the Paramount Value of Native Plants", *Landscape Architecture Magazine*, vol. 44, no. 4 (July 1954), pp. 200–08, p. 200.

In the face of his own death, British artist and filmmaker DEREK JARMAN (1942–94)
created a flourishing work of garden art in a location most people thought of as inhospitable:
rooted in the shingle on the south-east coast of England – next to a nuclear power plant.
by Nina Steinmüller, Photography by Howard Sooley

"My garden's boundaries are the horizon."

In 1986, Derek Jarman was touring southern England with a Super 8 camera, collecting footage for an experimental film that he would later call *The Garden*, when he called in at Dungeness in Kent. There, he, along with his companion Keith Collins and the actress Tilda Swinton, chanced upon Prospect Cottage, a small fisherman's hut that was up for sale. The smitten Jarman bought the house, moved in, and began creating a garden. But in December of that same year, he was diagnosed with HIV, and so the garden came to symbolize a life lived in the face of death.

At the time of moving to Dungeness, Jarman, who was born in 1942, was already internationally acclaimed as an artist and was becoming even more so for his films. Although he had shown an interest in art as

a child, his father initially prevented him from studying Art at university; on leaving school Jarman first went to King's College London to study English Literature, History, and the History of Art, and only after completing that degree did he enrol at Slade School of Fine Art, which he graduated from in 1968. While Jarman worked mainly as a painter – his works were shown in numerous exhibitions – he also designed stage sets and costumes, published poems, and in the late 1960s began experimenting with Super 8. Films such as *Sebastiane*, *Caravaggio*, and *Edward II* later made him world-famous. His was a radical and poetic cinematography that developed as a distinctive idiom of unforgettable images and dreamlike sequences unfurling a homosexual view

→ Derek Jarman's
Garden at Prospect
Cottage, Dungeness,
Kent, 1993

126

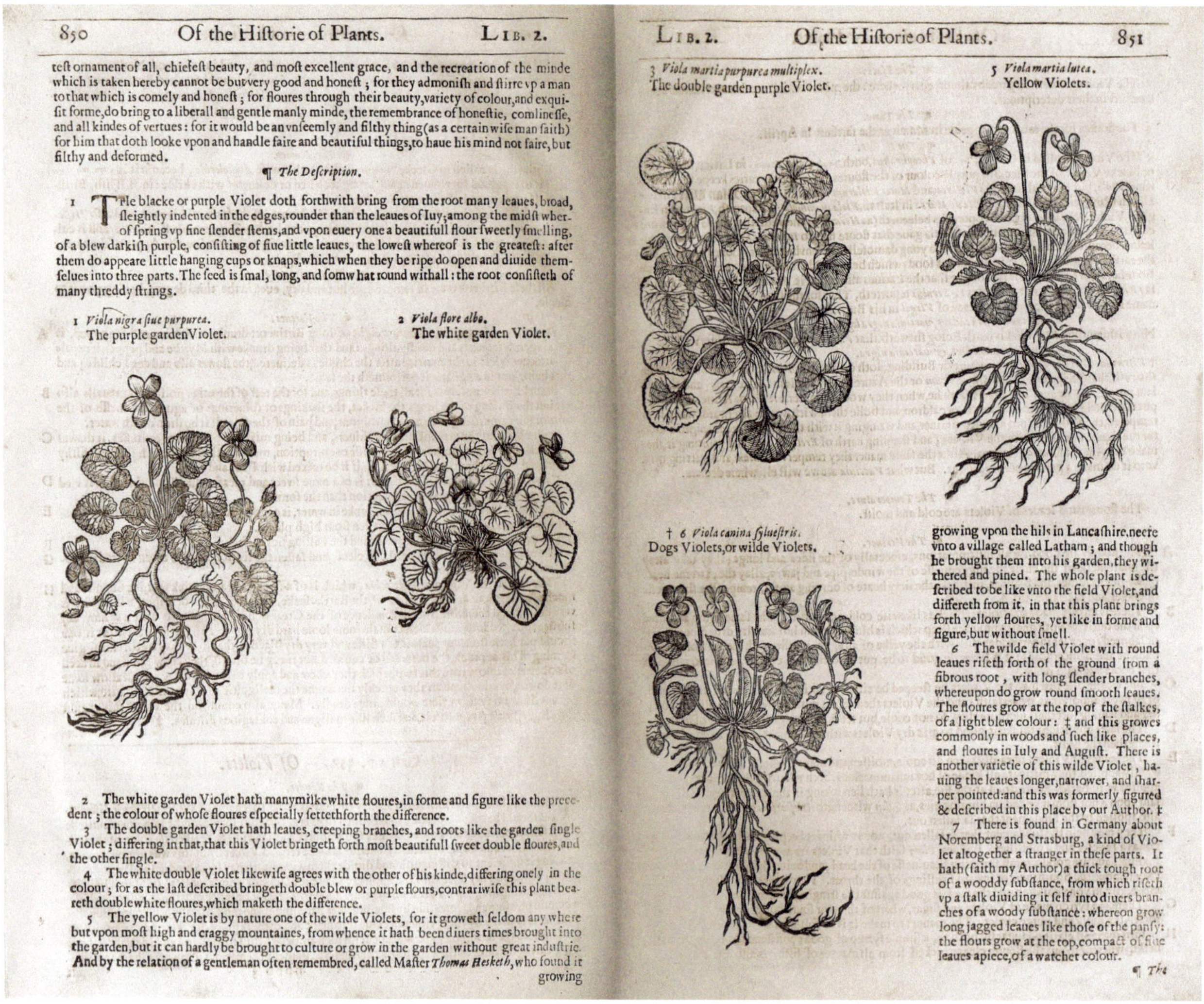

test ornament of all, chiefest beauty, and most excellent grace, and the recreation of the minde which is taken hereby cannot be but very good and honest; for they admonish and stirre vp a man to that which is comely and honest; for floures through their beauty, variety of colour, and exquisit forme, do bring to a liberall and gentle manly minde, the remembrance of honestie, comlinesse, and all kindes of vertues: for it would be an vnseemly and filthy thing (as a certain wise man saith) for him that doth looke vpon and handle faire and beautiful things, to haue his mind not faire, but filthy and deformed.

¶ *The Description.*

1 THe blacke or purple Violet doth forthwith bring from the root many leaues, broad, sleightly indented in the edges, rounder than the leaues of Iuy: among the midst wherof spring vp fine slender stems, and vpon euery one a beautifull flour sweetly smelling, of a blew darkish purple, consisting of fiue little leaues, the lowest whereof is the greatest: after them do appeare little hanging cups or knaps, which when they be ripe do open and diuide themselues into three parts. The seed is smal, long, and somwhat round withall: the root consisteth of many threddy strings.

1 Viola nigra siue purpurea.
The purple garden Violet.

2 Viola flore alba.
The white garden Violet.

2 The white garden Violet hath manymilke white floures, in forme and figure like the precedent; the colour of whose floures especially setteth forth the difference.
3 The double garden Violet hath leaues, creeping branches, and roots like the garden single Violet; differing in that, that this Violet bringeth forth most beautifull sweet double floures, and the other single.
4 The white double Violet likewise agrees with the other of his kinde, differing onely in the colour; for as the last described bringeth double blew or purple floures, contrariwise this plant beareth double white floures, which maketh the difference.
5 The yellow Violet is by nature one of the wilde Violets, for it groweth seldom any where but vpon most high and craggy mountaines, from whence it hath been diuers times brought into the garden, but it can hardly be brought to culture or grow in the garden without great industrie. And by the relation of a gentleman often remembred, called Master *Thomas Hesketh*, who found it growing

3 Viola martia purpurea multiplex.
The double garden purple Violet.

5 Viola martia lutea.
Yellow Violets.

† *6 Viola canina syluestris.*
Dogs Violets, or wilde Violets.

growing vpon the hils in Lancashire, neere vnto a village called Latham; and though he brought them into his garden, they withered and pined. The whole plant is described to be like vnto the field Violet, and differeth from it, in that this plant brings forth yellow floures, yet like in forme and figure, but without smell.
6 The wilde field Violet with round leaues riseth forth of the ground from a fibrous root, with long slender branches, whereupon do grow round smooth leaues. The floures grow at the top of the stalkes, of a light blew colour: ‡ and this growes commonly in woods and such like places, and floures in Iuly and August. There is another varietie of this wilde Violet, hauing the leaues longer, narrower, and sharper pointed: and this was formerly figured & described in this place by our Author. ‡
7 There is found in Germany about Noremberg and Strasburg, a kind of Violet altogether a stranger in these parts. It hath (saith my Author) a thick tough root of a wooddy substance, from which riseth vp a stalk diuiding it self into diuers branches of a woody substance: whereon grow long jagged leaues like those of the pansy: the floures grow at the top, compact of fiue leaues apiece, of a watchet colour.
¶ The

Paradise haunts gardens,
and some gardens are paradises.
Mine is one of them.[1]

← Pages on the violet
from John Gerard, *The
Herball, or, Generall
historie of plantes*, 1636,
cited by Derek Jarman
in his diary

↙ Seakale in spring in
Derek Jarman's garden
at Prospect Cottage,
1993

→ Derek Jarman, in
his garden at Prospect
Cottage, 1993

of history. After being diagnosed with HIV, Jarman's sexuality, which had been until then one influence among many, became the driving force of his art.[2]

Gardening had been part of Jarman's life since the age of four when his parents gave him the book *Beautiful Flowers and How to Grow Them*. Yet, cultivating plants in Dungeness, close to the coast where the ground was stony and the soil barren, seemed impossible at first. After initial attempts to grow roses failed dismally, he began looking for native species to plant which had already adapted to the raw conditions – above all, the constant wind and little rainfall. What really thrives at Dungeness is seakale (*Crambe maritima*), gorse, and poppies.

Jarman later became preoccupied with the cultural history of every plant in his garden. He collected old herbals from which he could read up about the symbolism, historical significance, and medicinal use of the plants and those that he found on his walks. He copied whole sections of the texts into his journals.

The English garden design legend Gertrude Jekyll was one of his most important influences, and he was acquainted with Beth Chatto, the plantswoman and author who became famous for her gravel garden. After Chatto visited Jarman in Dungeness in 1990, they continued corresponding by letter. These letters tell of their mutual esteem.[3]

> *The stones, especially the circles, remind*
> *me of dolmens, standing stones. They*
> *have the same mysterious power to attract.*[4]

The substrate of Jarman's garden is shingle worn smooth by the sea, which he arranged into large circles and squares at the front of his house. He created these geometric patterns by lining up flint and pebbles of various colours and sizes, and then interspersed them with bushes and shrubs. Behind the house lie small sculptural islands made of driftwood, rusty metal objects, and ornamental

rows of pebbles, none of which follow any particular order. These *objets trouvés* are paired with poppies, lavender cotton, foxglove, thistle, lathyrus, and flax. And while the garden has been thought of as a walled patch of nature shaped by human hand since the Middle Ages, Jarman's garden is the exact opposite: having no boundaries it draws our gaze to the desolate landscape beyond and all the way to the horizon, where the nuclear power plant looms like a blinking UFO.

It was in Dungeness that Jarman went on to shoot most of the footage for *The Garden*, a homoerotic version of the story of the Passion. Jarman's cottage provided him and his friends with a safe haven from the social stigma towards gays that had become especially pronounced during the AIDS crisis. Deeply shaken by the many deaths in his own circle of friends, Prospect Cottage became all the more a refuge, whose garden was to have the cathartic effect of awakening new life.

↑ Jarman working in his garden, 1991

↗ Poppies and rusty finds in the back garden, 1993

→ Planting plan and key in Jarman's *Garden Book I*, 1989

I walk in this garden
Holding the hands of dead friends
Old age came quickly for my frosted generation
Cold, cold, cold they died so silently
Did the forgotten generations scream?
Or go full of resignation
Quietly protesting innocence
Cold, cold, cold they died so silently
[...]
My gilly flowers, roses, violets blue
Sweet garden of vanished pleasures
Please come back next year
Cold, cold, cold I die so silently[5]

After Jarman's own death in 1994, Collins looked after the garden until he himself died in 2018. A fund-raising campaign by the British Art Fund in 2020 ensured that Prospect Cottage and its garden would be preserved for the future. Both house and garden can now be visited by prior appointment.

The photographer Howard Sooley first met Jarman in 1990 when he came to take the artist's portrait. The two men became close friends, and right to the end Sooley visited Jarman regularly in order to garden with him. His photographs of the artist and his garden are a unique testimony which capture the timeless, mystical atmosphere of Dungeness ●

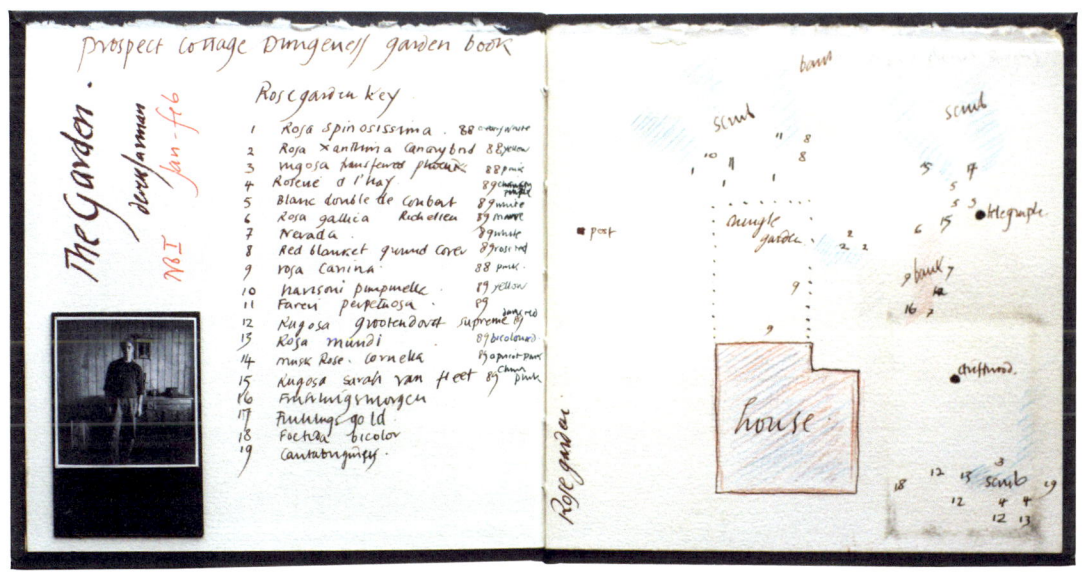

1   *Derek Jarman's Garden*, with photographs by Howard Sooley. London: Thames & Hudson, 1995, p. 40.

2   Roger Wollen, "Chronology", in *Derek Jarman: A Portrait*, with an introduction by Wollen. London: Thames & Hudson, 1996, p. 168.

3   Today the letters are kept in the Beth Chatto Archive at the Garden Museum in London. See also, "When Beth Met Derek", Garden Museum (3 July 2020), online: https://garden-museum.org.uk/when-beth-met-derek, accessed 21 November 2022.

4   *Derek Jarman's Garden*, with photographs by Howard Sooley. London: Thames & Hudson, 1995, p. 24.

5   Derek Jarman, *Modern Nature*. Woodstock, NY: Overlook Press, 1994, pp. 69–70.

A garden can form the basis of an entire company, as it does at cosmetics and natural medicine producer WELEDA founded in Switzerland in 1921. Here, business interests are subject to the finely calibrated cycles of nature – a generations-old model for the future as described by the head gardener of the company's own gardens.

by Astrid Sprenger

Gardening to the Cycles and Rhythms of Nature

The most wonderful thing about gardens is that they have neither a beginning nor an end, but are rather a never-ending story of growth and decay in harmony with the seasons. By tending the soil, sowing, pricking out and planting seedlings, weeding and watering, harvesting and collecting next year's seeds, composting and mulching, we as gardeners of an organic garden are an active and formative part of this cycle. Whatever we extract from the garden, we return to it in kind; the only limits to our creativity being those set by nature itself. We are learners our whole lives long and our dialogue with nature is our most important resource.

We operate six gardens all over the world: in Germany, the Netherlands, England, Brazil, Argentina, and Aotearoa New Zealand. The bulk of the medicinal plants used to make Weleda products are grown in these gardens. All our gardens are cultivated according to the principles of biodynamic farming, which entails viewing the garden as a single organism. Dispensing with artificial fertilizers and synthetic herbicides and pesticides, we design species-appropriate habitats for our plants and pay close attention to soil structure and biodiversity. This makes for a stable ecosystem that rewards us with sustainable and healthy yields. Even as we pursue our principal objective, which is to cultivate, harvest, and process medicinal plants of the very highest quality to promote human well-being, we are also creating ecosystems and gardens that are of benefit to the common good.

→ Astrid Sprenger at her place of work in Schwäbisch Gmünd, where she has been head of medicinal plant cultivation since 2022. Almost everything is done by hand here.

↘ The garden is
a complex organism
that keeps itself
healthy. Where
diversity can unfold,
there is stability.

The most wonderful thing about gardens is that
they have neither a beginning nor an end, but are
rather a never-ending story of growth and decay.

We grow more than 200 different plant varieties in our 23-hectare garden in Schwäbisch Gmünd. Every plant has its own set of demands regarding site, soil, water, and light. From some we harvest the flowers, leaves, or stalks, from others the fruits, bark, or roots. It is therefore vital that we replicate each plant's ideal natural habitat as closely as possible. Hart's tongue fern, for example, grows in forests, which is why we grow it in soil covered in leaf mould under a knotweed tunnel. In summer, the knotweed provides shade, while in autumn it sheds its leaves and lets in more light. The result is a sylvan microclimate even in high summer. In other words, we pay more attention to getting the habitat right than to the plants that are to grow there.

Our top priority is the right kind of soil. Because they are constantly interacting with the atmo-, hydro-, litho-, and biospheres, soils are highly complex ecosystems in their own right. But only healthy soils can fulfil such key functions as the sequestration of water, nutrients, and carbon, and only healthy soils produce healthy plants. Healthy soils are sensitive and teeming with life. In fact, there are more living organisms in a handful of soil than there are human beings on the whole planet! We therefore try to minimize our interventions and do only whatever is necessary to promote healthy soil. Hence, the several hundred cubic metres of compost that we produce every year. By feeding our living soils with humus – that is, with decayed organic matter from our compost – we help fix carbon in the soil. The root system of the plants, aided by the bacteria and fungi in the soil, can then draw all the nutrients and water it needs.

We practise crop rotation and wherever possible sow each bed with something different every year so as not to deplete the soil of any one nutrient (or group of nutrients). We sow green manure after every harvest and grow rows of flowers between the beds of medicinal plants. Not only does green manure shield the soil from physical forces such as torrential rain and extreme temperatures, but it also fixes nitrogen from the air and keeps the soil airy and loose. It is also a natural way of replenishing the soil's nutrients and other substances conducive to healthy biodiversity. The green manure and rows of flowers provide food for countless insects and birds and so promote biodiversity above ground too. Our fields also feature miles of hedgerow full of native trees and bushes as well as agroforestry plantations. These supply a habitat and a food source for countless

↖ Weleda gardens are biodiverse places that offer habitats for as many plants and animals as possible.

↑ Vegetable oils are a valuable basis for Weleda natural cosmetics.

We are helping to improve soil quality, enhance biodiversity, conserve water, and create carbon sinks.

small animals, insects, and birds. By cultivating the living conditions that beneficial fauna need to thrive, we obviate the need for pesticides, in other words.

Many of the challenges that the world is now facing are a result of our failure to appreciate just how complex the interdependencies of nature are. Since the middle of the last century, conventional agriculture has by and large followed classic economic principles. To promote profits and growth, nature has been standardized globally and stripped of biodiversity. This has resulted in vast monocultures that yield bumper crops only if they are regularly sprayed with artificial fertilizers and agrochemicals to keep them free of pests and weeds. The long-term negative impacts of these methods are as glaringly obvious as they are devastating: eroded, barren soils utterly lacking in microorganisms, water shortages, contaminated groundwater, greenhouse gas emissions, and dwindling biodiversity. Ultimately, what we are doing is destroying our own

human habitat. Conventional thinking lacks the farsighted, systemic view of nature's delicate balance of complex interdependencies.

Through our work in Weleda's gardens we are helping to improve soil quality; enhance biodiversity; conserve water, carbon, and nutrients; secure a plentiful supply of productive seed stock; and harvest large crops. Our gardens prove that humans can also have a beneficial influence on nature and live sustainably both with it and from it. So perhaps our most important message in this time of crisis is this: we *can* build a good future, at least as long as we begin working with nature and not against it. And where better to reassure ourselves of this truth than in the garden? ●

↖ Plant processing in the Weleda Production Plant, Schwäbisch Gmünd, late 1920s

→ Cultivation of calendula plants, apothecary garden, Schwäbisch Gmünd, Germany, 2016

For the past 30 years JAMAICA KINCAID'S Vermont garden
has been more than a place of work and pleasure. For
Kincaid (b. 1949), it has also been a springboard to interrogate aspects
of colonial history, cultural appropriation, and displacement.

The Disturbances of the Garden – In the garden, one performs the act of possessing

My obsession with the garden and the events that take place in it began before I was familiar with that entity called consciousness. My mother taught me to read when I was very young, and she did this without telling me that there was something called the alphabet. I became familiar with words as if they were all wholly themselves, each one a world by itself, intact and self-contained, and able to be joined to other words if they wished to or if someone like me wanted them to. The book she taught me to read from was a biography of Louis Pasteur, the person she told me was responsible for her boiling the milk I drank daily, making sure that it would not infect me with something called tuberculosis. I never got tuberculosis, but I did get typhoid fever, whooping cough, measles, and persistent cases of hookworm and long worms. I was a "sickly child". Much of the love I remember receiving from my mother came during the times I was sick. I have such a lovely memory of her hovering over me with cups of barley water (that was for the measles) and giving me cups of tea made from herbs (bush) that she had gone out and gathered and steeped slowly (that was for the whooping cough). For the typhoid fever, she took me to the hospital, the children's ward, but she visited me twice a day and brought me fresh juice that she had squeezed or grated from fruits or vegetables, because she was certain that the hospital would not provide me with proper nourishment. And so there

I was, a sickly child who could read but had no sense of consciousness, had no idea of how to understand and so make sense of the world into which she was born, a world that was always full of a yellow sun, green trees, a blue sea, and Black people.

My mother was a gardener, and in her garden it was as if Vertumnus and Pomona had become one: she would find something growing in the wilds of her native island (Dominica) or the island on which she lived and gave birth to me (Antigua), and if it pleased her, or if it was in fruit and the taste of the fruit delighted her, she took a cutting of it (really she just broke off a shoot with her bare hands) or the seed (separating it from its pulpy substance and collecting it in her beautiful pink mouth) and brought it into her own garden and tended to it in a careless, everyday way, as if it were in the wild forest, or in the garden of a regal palace. The woods: the garden. For her, the wild and the cultivated were equal and yet separate, together and apart. This wasn't as clear to me then as I am stating it here. I had only just learned to read and the world outside a book I did not yet know how to reconcile.

The only book available to me, a book I was allowed to read all by myself without anyone paying attention to me, was the King James Version of the Bible. There's no need for me to go into the troubles with the King James Version of the Bible here, but when

→ Jamaica Kincaid
in the garden at here
home in Vermont

The day that Christopher Columbus set sail from Spain, later having a fatal encounter with the Indigenous people he met in the "West Indies", the world of the garden changed.

I encountered the first book, the Book of Genesis, I immediately understood it to be a book for children. A person, I came to understand much later, exists in the kingdom of children no matter how old the person is; even Methuselah, I came to see, was a child. But never mind that, it was the creation story that was so compelling to me, especially the constant refrain "And God saw that it was good". The God in the Book of Genesis made things, and at the end of each day he saw that they were good. But, I wondered, for something to be good would there not have to be something that was not good, or not as good? That was a problem, though I didn't bother myself with it at the time, mainly because I didn't know how to, and also because the story had an inexorableness to it: rolling on from one thing to another without a pause until, by the end of six days, there were a man and a woman made in God's image, there were fish in the sea and animals creeping on land and birds flying in the air and plants growing, and God found it all good, because here we are.

It was in the week after this creation, on the eighth day, that the trouble began: loneliness set in. And so God made a garden, dividing it into four quarters by running water through it (the classic quadrilinear style that is still a standard in garden design) and placing borders, the borders being the eternal good and evil: the Tree of Life and the Tree of Knowledge. One tree was to be partaken of, the other forbidden. I have since come to see that in the garden itself, throughout human association with it, the Edenic plan works in the same way: the Tree of Life is agriculture and the Tree of Knowledge is horticulture. We cultivate food, and when there is a surplus of it, producing wealth, we cultivate the spaces of contemplation, a garden of plants not necessary for physical survival. The awareness of that fact is what gives the garden its special, powerful place in our lives and our imaginations. The Tree of Knowledge holds unknown and therefore dangerous possibilities; the Tree of Life is eternally necessary, and the Tree of Knowledge is deeply and divinely dependent on it. This is not a new thought

for me. I could see it in my mother's relationship to the things she grew, the kind of godlike domination she would display over them. She, I remember, didn't make such fine distinctions, she only moved the plants around when they pleased her and destroyed them when they fell out of favour.

It is no surprise to me that my affection for the garden, including its most disturbing attributes, its most violent implications and associations, is intertwined with my mother. As a child, I did not know myself or the world I inhabited without her. She is the person who gave me and taught me the Word.

But where is the garden and where am I in it? This memory of growing things, anything, outside not inside, remained in my memory – or whatever we call that haunting, invisible wisp that is steadily part of our being – and wherever I lived in my young years, in New York City in particular, I planted: marigolds, portulaca, herbs for cooking, petunias, and other things that were familiar to me, all reminding me of my mother, the place I came from. Those first plants were in pots and lived on the roof of a diner that served only breakfast and lunch, in a dilapidated building at 284 Hudson Street, whose ownership was uncertain, which is the fate of us all. Ownership of ourselves and of the ground on which we walk, ownership of the other beings with whom we share this and see that it is good, and ownership of the vegetable kingdom are all uncertain, too. Nevertheless, in the garden, we perform the act of possessing. To name is to possess; possessing is the original violation bequeathed to Adam and his equal companion in creation, Eve, by their creator. It is their transgression in disregarding his command that leads him not only to cast them into the wilderness, the unknown, but also to cast out the other possession that he designed with great clarity and determination and purpose: the garden! For me, the story of the garden in Genesis is a way of understanding my garden obsession.

The appearance of the garden in our everyday life is so accepted that we embrace its presence as therapeutic. Some people say that weeding is a form of

Impressions from Jamaica Kincaid's Instagram account @virtuouspomona

↗ "An Antiguan garden in Vermont by way of Mexico, Southern India, Southern Africa." 24 September 2021

→ "Tallest lily of all the indigenous North American lilies: *Lilium superbum*. To see it in a garden is a triumph; to come on it in a meadow is to realize there is no such thing as Triumph!" 3 August 2021

comfort and of settling into misery or happiness. The garden makes managing an excess of feelings – good feelings, bad feelings – rewarding in some way that I can never quite understand. The garden is a heap of disturbance, and it may be that my particular history, the history I share with millions of people, begins with our ancestors' violent removal from an Eden. The regions of Africa from which they came would have been Eden-like, and the horror that met them in that "New World" could certainly be seen as the Fall. Your home, the place you are from, is always Eden, the place where even imperfections were perfect, and everything that happened after that beginning interrupted your Paradise.

On August 3, 1492 – the day that Christopher Columbus set sail from Spain, later having a fatal encounter with the Indigenous people he met in the "West Indies" – the world of the garden changed. That endeavour, to me, anyway, is the way the world we now live in began; it not only affected the domestic life of Europeans (where did the people in a Rembrandt painting get all that stuff they are piling on?) but suddenly they were well-off enough to be interested in more than sustenance, or the Tree of Life (agriculture); they could now be interested in cultivating the fruits of the Tree of Knowledge (horticulture).

Suddenly, the conquerors could do more than feed themselves; they could also see and desire things that were of no use apart from the pleasure that they produced. When Cortés saw Montezuma's garden, a garden that incorporated a lake on which the capital of Mexico now sits, he didn't mention the profusion of exotic flowers that we now grow with ease in our own gardens (dahlias, zinnias, marigolds).

The garden figures prominently in the era of conquest, starting with Captain Cook's voyage to regions that we now know as Australia, New Zealand, New Guinea, and Tahiti; its aim, ostensibly, to observe the rare event of the transit of Venus. On this trip, in 1768, the first of Cook's three voyages around the world, he brought with him the botanist Joseph Banks and also Daniel Charles Solander, a student of Carolus Linnaeus. The two took careful notes on everything they saw. Banks decided that the breadfruit of the Pacific isles would make a good food for slaves on British-owned islands in the West Indies; the slaveholders were concerned with the amount of time it took the enslaved people to grow food to sustain themselves, and breadfruit grew with little cultivation. And so the Pacific Islands came to the West Indies. Banks also introduced the cultivation of tea (*Camellia sinensis*) to India.

Then there is Lewis and Clark's expedition from the Mississippi River to the Pacific Northwest. On that adventure, which was authorized by President Thomas Jefferson and was inspired by Cook's scientific and commercial interests, the explorers listed numerous plant species that were unknown to John Bartram, botanist to King George III, who ruled the United States when it was still a colony. Bartram's son, William, a fellow botanist, later wrote a book about his own explorations, which is said to have influenced Wordsworth, Coleridge, and other English Romantic poets.

There now, look at that: I am meaning to show how I came to seek the garden in corners of the world far away from where I make one, and I have got lost in thickets of words. It was after I started to put seeds in the ground and noticed that sometimes nothing happened that I reached for a book. The first ones I read were about how to make a perennial border or how to get the best out of annuals – the kind of books for people who want to increase the value of their home – but these books were so boring. I found an old magazine meant to help white ladies manage their domestic lives in the 1950s much more interesting (that kind of magazine, along with a copy of *Mrs. Beeton's Book of Household Management*, is worthy of a day spent in bed while the sun is shining its brightest outside). But where did plants, annual and perennial, pristinely set out in something called a border, and arranged sometimes according to colour and sometimes according to height, come from? Those books had no answer for me. So one book led to another, and before long I had acquired (and read) so many books that it put a strain on my family's budget. Resentment, a not unfamiliar feeling relating to the garden, set in.

I began to refer to plants by their Latin names, and this so irritated my editor at this magazine [*New Yorker*, Veronica Geng] that she made me promise

that I would never learn the Latin name of another plant. I loved her very much, and so I promised that I would never do such a thing, but I did continue to learn the Latin names of plants and never told her. Betrayal, another feature of any garden.

How did plants get their names? I looked to Linnaeus, who, it turned out, liked to name plants after people whose character they resembled. Mischievous, yes, but not too different from the doctrine of signatures, which attempted to cure diseases by using plants that resembled the diseased part of the body. I was thinking about this one day, stooped over and admiring a colony of *Jeffersonia diphylla*, whose common name is twinleaf. *Jeffersonia diphylla* is a short woodland herbaceous ephemeral whose leaf is perforated at the base so that it often looks like a luna moth, but the two leaflets are not identical at the margins, and each leaf is not evenly divided: the margins undulate, and one leaflet is a little bigger than the other. But isn't Thomas Jefferson, the gardener, the liberty lover and slaveowner, often described as divided, and isn't it appropriate that a plant such as the twinleaf is named for him? The name was bestowed by one of his contemporaries, Benjamin Smith Barton, who

Impressions from Jamaica Kincaid's Instagram account @virtuouspomona

↙ "Marigolds and Dahlias. Two of the gems the Spanish thief and murderer Hernán Cortés found when he met Montezuma." 3 August 2021

↓ "Hot peppers and sweet potatoes." 11 October 2021

How did plants get their names? I looked to Linnaeus, who, it turned out, liked to name plants after people whose character they resembled.

perhaps guessed at his true character. It was through this plant that I became interested in Thomas Jefferson. I have read much of what he wrote and have firm opinions about him, including that his book *Notes on the State of Virginia* is a creation story.

It was only a matter of time before I stumbled on the plant hunters, although this inevitability was not clear to me at all. Look at me: my historical reality, my ancestral memory, which is so deeply embedded that I think the whole world understands me before I even open my mouth. A big mistake, but a mistake not big enough for me to have learned anything from it. The plant hunters are the descendants of people and ideas that used to hunt people like me.

The first one I met, in a book, of course, was Frank Smythe. No one had ever made me think that finding a new primrose – or a new flower of any kind – was as special as finding a new island in the Caribbean Sea when I thought I was going to China to meet the Great Khan. A new primrose is more special than meeting any conqueror. But Smythe gave me more than that. I noticed, when reading his accounts, that he was always going off on little side journeys to climb some snow-covered protuberance not so far away, and then days later returning with a story of failure or success at reaching or not reaching the peak, and that by the way he had found some beauty of the vegetable kingdom on the banks of a hidden stream which would be new to every benighted soul in England. But his other gift to me was the pleasure to be had in going to see a plant that I might love or not, growing somewhere far away. It was in his writing that I found the distance between the garden I was looking at and the garden in the wilderness, the garden cast out of its Eden which created a longing in me, the notion of "to go and to see". Go see!

I end where I began: reading – learning to read and reading books, the words a form of food, a form of life, and then knowledge. But also my mother. I don't know exactly how old I was when she taught me to read, but I can say for certain that by the time I was three and a half I could read properly. This reading of mine so interfered with her own time to read that she enrolled me in school; but you could be enrolled in school only if you were five years old, and so she told me to remember to say, if asked, that I was five. My first performance as a writer of fiction? No, not that at all. Perhaps this: the first time I was asked who I was. And who am I? In an ideal world, a world in which the Tree of Life and the Tree of Knowledge stand before me, before all of us, we ask, who am I? Among the many of us not given a chance to answer is the woman in the library in St. John's, Antigua, two large rooms above the Treasury Department, a building that was steps away from the customs office and the wharf where things coming and going lay. On that wharf worked a stevedore who loaded onto ships bags of raw sugar en route to England, to be refined into white sugar, which was so expensive that we, in my family, had it only on Sundays, as a special treat. I did not know of the stevedore, the lover of this woman who would not allow her children to have much white sugar because, somewhere in the world of Dr. Pasteur and his cohort, they had come to all sorts of conclusions about diseases and their relationships to food – beriberi was a disease my mother succeeded in saving me from suffering. Her name was Annie Victoria Richardson Drew, and she was born in a village in Dominica, British West Indies ●

Case Study #6
PIET OUDOLF (b. 1944) is one of the most renowned garden designers working today. Celebrated for his elaborate artistic planning, his designs embrace the beauty of all the seasons.
by Hanno Rauterberg

Can Art Be Nature?

"Society sees itself reflected in its gardens" – a statement like this is typical of Piet Oudolf. And he makes it almost in passing, as if it were a truism that is surely blindingly obvious to everyone, and especially to those at the DIY store hunting for a digital irrigation system or some window box flowers for the balcony; the same store which just recently started stocking large wooden boxes that supposedly make growing vegetables both faster and easier. Is that society, too – this crating of vegetation, this casketing of the earth and the fruits of the earth along with it?

Oudolf is not a sociologist, nor is he a philosopher. He is an artist, some say. Certainly, his name has been bandied about in the art world for some time now, and he has designed many a luxuriant garden for American museums. Hauser & Wirth gallery in England also had him stage a performance of perennials and grasses for its Somerset location. And in Weil am Rhein, the garden now growing alongside the Design Museum of the furniture manufacturer Vitra is an open invitation to an aesthetic experience without parallel. For here, nature becomes art, and art nature. Does society see itself reflected here, too?

Oudolf is certainly proud, if a little surprised that someone like him, a garden and landscape designer who for decades was only known in horticultural circles, is suddenly being hailed by art and design magazines as an avant-garde artist. And he can scarcely cope

with the flood of enquiries that he receives. The perpetual extravagance of the art world seems rather remote out here in Hummelo, a poor, industrialized landscape geared towards efficiency not far from the German-Dutch border, where Oudolf and his wife Anja converted an old farmhouse and where at this very minute a herd of black and white piebald ruminants are grazing nonchalantly in front of his studio window.

If Oudolf is an artist, then he is one who lives his ideas – a conceptual artist, in other words. He sits at his desk strewn with coloured crayons and tracing paper. A few squiggles of purple here, some small blue dots there: the sheets are full of mysterious patterns at once ebullient and entirely abstract – at least to the layman. To Oudolf, they are speaking signs. Because hidden behind each line is a three- or rather four-dimensional image of colour, shape, mood, scent, and even swishing sounds. Although "image" is the wrong term here, since what Oudolf actually creates is a multilayered, multivalent experience. Anyone who plants a garden, plants in time.

Oudolf says that before going to sleep he likes to immerse himself in this sensory world and review his drawings in his mind's eye. How might it look if the quivering dewdrop grass was interspersed with cream-pink allium globes? Is there enough contrast between their textures? How will the colour interact

→ Aerial view of the Oudolf Garten on the Vitra Campus, Weil am Rhein, Germany, designed in 2020

with some rippling rows of sage, and how will that combination change come August or October? And what will happen when the Eupatorium shoots up in the background? Will it then dominate the scene? Will the shimmering grass be able to hold its own? His head is teeming with a hundred of such combinations; Oudolf sees the plants bud, flower, wilt, and wither away. He looks on as the perennials and shrubs and grasses steadily mesh, and wonders whether the feebler are too feeble or the flamboyant are flamboyant enough.

People go into raptures over the naturalness of his planting, over the way order seems to elude its own strictures, becoming impulsive and wonderfully untamed. Unlike the ornamental borders invented and perfected by the English, Oudolf avoids all things forced and is instantly alienated by any claim to perfection. But naturalness? Wildness? "It's all controlled", says Oudolf, "all calculated". What looks like freedom, like gentle juxtapositions, turns out to be rigorously planned.

It is this common thread running through all of Oudolf's gardens that makes them so akin to art. Optical illusion is part of it, yet the illusory is also entirely real, even palpable. This strangely dualistic character is so compelling that for a moment we are inclined to believe that what is being revealed to us here is an alternative to our world of depleting biodiversity, melting ice caps, and the desertification of large parts of the land mass. Oudolf makes the environment a convironment. His gardens reset the balance and tell of a newly gained equilibrium.

But for all the romanticism, never once does he tend towards sentimentality – on the contrary.

Even his best-known design, that for the High Line in New York, shows how he can successfully turn the ruins of an abandoned railway viaduct into a botanical adventure, without hiding the remains of a bygone industrialism. By letting the rough and barren flourish, Oudolf stages an exciting clash of opposites. Even the ballast itself becomes fertile at his hands. And it is this transformative power, this honouring of a promise that even the dry remnants of the fossil fuel age may allow new shoots, which has made the High Line so incredibly popular.

The Vitra garden in Weil am Rhein does not refuse the presence of industry either, although it is easy to lose yourself here while strolling along its looping, twisting paths, all the while marvelling at the tall burnet whose pink calyxes seem to capture sunlight even on overcast days, delighting in the yarrow whose acidic yellow is enough to make your eyes sting, and swaying with the autumn moor grass between the little blue islands of crane's-bill. No matter how transfixed we are by this symphony of so many headstrong players, a second reality is bound to impress itself on our consciousness: traffic noise from the busy roads and heavily used railway line nearby tell us that here in this garden the peace of the natural world is miles away and our growth-obsessed society very close.

Oudolf gifts us whole eco-neighbourhoods. His gardens are at once panopticons and places of

← Detail of a planted area in the Oudolf Garten on the Vitra Campus

→ Piet Oudolf, Planting design sketch for the Oudolf Garten on the Vitra Campus

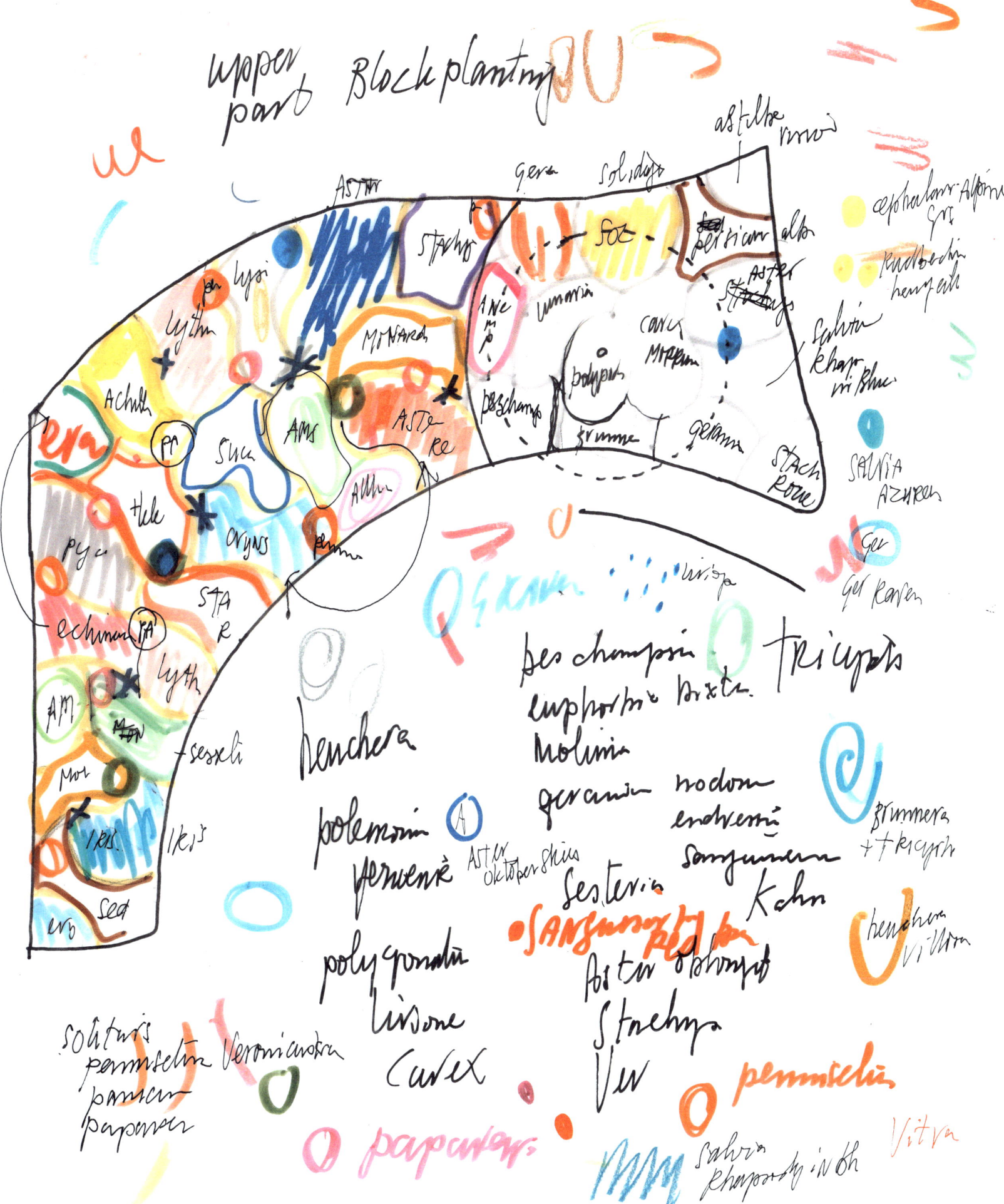

upper part Block planting

His gardens enable us to experience
a different relationship between
humanity and the world we inhabit.

contemplation, ideally suited for pondering those questions of life that amidst so many flowers seem almost to be growing up to greet us. Oudolf speaks of how bonding with plants allows us to connect to our true selves, to form a bond between our inner and outer nature. And his gardens really do enable us to experience a relationship between humanity and the world we inhabit rather differently, which might even be called "beautiful codependency". For while Oudolf's nature would be nothing without those avid horticulturalists who have crossed and recrossed ordinary meadow plants to produce something quite artificial, the gardener is not entirely free. He needs sunlight, he needs rain, he needs the microorganisms in the soil. Because without nature, culture cannot thrive. This dependence is at once trying and beautiful, in that it makes the garden, as tamed as it may be, a place of unpredictability. "The expectation", says Oudolf, "is that everything may turn out quite differently". It pleases him, the autonomous life of plants, and that for all his planning and encyclopaedic knowledge they elude his control. Yet, he is no friend of laissez-faire, of allowing everything to proliferate freely, but nor does he subscribe to the idea common in the world of engineering everything.

Oudolf thinks in cyclical processes. For him, there can be no growth without decay. His favourite months are September and October, which is when the vitality flips to melancholy – a transformation he loves to stage. He leaves the dead heads standing, savours their slow disintegration and the moment they keel over, when the last seed pods become encased in hoar frost. And he loves to watch the greenness become ever greyer, browner, deader. Reason enough to wait until spring before clearing it away, when the shoots and buds begin bursting forth all over again.

The yearning for balance, then, of which Oudolf's gardens speak, is never truly fulfilled. Because the balance he seeks must be attempted anew every year. And those who wish to know how that happens have no choice but to return again and again to Oudolf's great metaphor for society in Weil am Rhein. After all, a garden lives from its interactions and from the constantly shifting relationships between nature and culture. What thrives best here is the unexpected ●

This text is a revised version of the article "Der Traum vom wilden Leben", published in: *Die Zeit*, no. 26 (24 June 2021). Used by permission of Hanno Rauterberg and *Die Zeit*.

At first glance, contemporary digital culture and gardening may seem to have little in common. Inspired by the popular computer game *Age of Empires*, artist ZHENG GUOGU (b. 1970) combined aspects of Chinese tradition with pop culture to create a sprawling garden in his home town of Yangjiang.
Interview by Yujia Bian

"It's okay to borrow from popular culture."

Zheng Guogu's Liao Garden may never be complete. His 20,000-square-metre site is a constantly changing artwork, that in keeping with the practice of the artist revolves around the impact of digital technologies and Western culture on Chinese tradition and contemporary life.

Born in 1970 in Yangjiang, in the Chinese Guangdong Province, where he still lives and works, Guogu graduated from the printmaking department at the Guangzhou Academy of Fine Arts in 1992. Today, he works across a range of media. His work has been exhibited internationally, including at the Venice Biennale, Guggenheim Museum, and Museum of Modern Art, New York.

In 2000, Guogu purchased land in his hometown, and five years later he started constructing an enormous garden based on a hugely popular video game *Age of Empires*, in which players occupy land to create new worlds. Guogu's garden is also rooted in Chinese philosophy and following his artistic experimentations combines complex spatial modalities with social relations. For Liao Garden, Guogu melded the work of an artist with that of a farmer or gardener. He dug canals and made mountains, planted trees and built houses, just like in the game; the garden now includes a golden pyramid and a welcoming atrium. The work is conceptual, but also personal – the artist's imprints lie deep in the plot of land. Yujia Bian spoke to Zheng Guogu about the 20-year endeavour that is Liao Garden.

→ Zheng Guogu, Liao Garden, Yangjiang, designed from 2000

↑ Zheng Guogu in
Liao Garden, 2015

↗ Zheng Guogu,
Sketch for the Museum
of Wonder, 2005

How did Liao Garden come about?

In 2000, I was searching for a large studio. I wanted it to be a garden in style, with different levels. I found land that was supplied with water, electricity, and road access, and I discovered it had a hilly orchard full of lychees. It made me think that this kind of sloped landscaping should be reflected in the built structures as well as in the garden. The idea is a bit like the gardens of the Imperial Era that have both mountains and forests within them.

You turned this orchard into a garden. What would you say is the difference between the two?

A garden imitates the diversity of plants found in the wild on an artificial level. Plant species are diverse in nature, not homogenous – you only need to look at the ecology of nature. The diversity is seen in the competition between species and in their cross-pollination. If you see an orchard, it must have been planted; it is the opposite of variegated. So, I diversified the plant life in Liao Garden to recover that mix of species, and now it has the look of a garden.

Initially the garden was called "Age of Empires". Why?

Back then I was playing a video game called *Age of Empires*. I liked the game because you can expand your territories. Starting from peasantry and farming, slowly you can build up an army or construct a castle. Eventually, when your strength reaches a certain point, you can build something called a "Wonder of the World". These are different monuments from the histories of different nations. That is to say, in Egypt, it is a pyramid; in China, it may be the Temple of Heaven. I thought I could realize these virtual things in reality. Just like Pop art, it's okay to borrow from what is common knowledge. There were about 80 million people playing *Age of Empires* in China at the time, so I thought why not import the game's elements into the idea of the Chinese garden, but not a Ming or Qing Dynasty garden, or a Suzhou garden, or like the Li Garden in Kaiping, Guangdong, but a contemporary one. So, we made a garden of *this* era.

There were about 80 million people playing
Age of Empires, I imported the game's
elements into the idea of the Chinese garden.

↗ A screenshot
from the video game
Age of Empires

↑ → Museum in the
Wind, Liao Garden, 2017
(started from 2005)

One should keep the spirit of tradition and not lose the spirit of the garden.

There is also a "Museum in the Wind".

The Museum in the Wind is a series of open layers connected by staircases, but it was originally supposed to be an art museum. We need to do projects that are culture-related to be given legal status by the local government. So, we proposed to make a museum. But we had nothing to exhibit! That's why we have a museum composed solely of levels, pathways, and stairs. The experience of being in the museum differs dramatically from floor to floor by the way one feels the wind blowing in from the mountains. The museum only exhibits the wind, which is different every day. There are exhibitions if there is wind and there are none if there is no wind. That's the concept.

Once I saw an Indian yoga centre in the form of a circle. Most yoga centres in India are round. And because the central complex where I work and meet with guests had a roof, I decided to cut a circle of about twelve metres in diameter into it. I then created a garden in the atrium, replete with flowing water. The life of water is in concert with the heart. Your heart also flows like water. That is to say, when you go to the central building, you hear the sound of flowing water and you see the water flowing outside.

There is also a frangipani garden, which is a plant rarely seen in Chinese gardens. And a landscape feature called *tianchi*, whose contours are like terraced fields and each layer has a pool of water, and there are over a dozen layers.

It seems as if you have created Liao Garden as your particular world or universe. Can we understand it in this way?

The scope of Liao Garden is pretty much like those Chinese private gardens of the past. For example, the classical Suzhou gardens that encircle nature – you are inside and couldn't care less about the outside. One should keep the spirit of tradition and not lose the spirit of the garden. And just like a person, the connection with the sky should not be severed. Although many landscape architects fail to take sight of this connection or if it is broken. This is the most important thing to consider if you want to maintain a relationship with the laws of the universe ●

The introduction to this interview is based on excerpts originating in *From Semiotics to Energetics*, vol. 1: *Zheng Guogu: The Everlasting Garden*. Guangzhou: The Pavilion, 2021. Courtesy of The Pavilion. Interview originally published in *Spike* art magazine, no. 65 (Autumn 2020), pp. 28–31.

In Kuala Lumpur, one of the world's most densely built-up megacities, the civic engagement and community spirit of a group of citizens led to the KEBUN-KEBUN BANGSAR initiative, which repurposed wasteland into a thriving garden. And this is hopefully just the beginning, says one of the initiators, NG SEK SAN.
Interview by Viviane Stappmanns

"All of our work is about connecting."

Kuala Lumpur-based landscape architect Ng Sek San retired from professional practice in 2013, but he did not stop working; after 30 years of delivering projects for commercial clients and government agencies, he started to serve his community. Kebun-Kebun Bangsar, a neighbourhood garden project initiated post the general election campaign of that year, was the first of its kind in the Malaysian capital. Situated on eight acres of land in central Kuala Lumpur, the garden was conceived as an act of resistance to top-down planning of public spaces with little regard for community needs and participation. Operating since 2017 with temporary permissions – recently extended after protests against potential closure

proved successful – the garden is not only a haven for families and volunteers, but embedded in a larger system of local supply and waste management. It turns the kitchen scraps of the entire neighbourhood, including those of shopping complexes, into compost, and supplies food to families in the area. Apart from being an asset to the community, the garden is a model for biophilic (increased connectivity to the natural environment) and decentralized design practice, which Ng Sek San hopes will be applied elsewhere in the city, where population density is high and planning for adequate green space still lags behind the rapid urbanization that has prevailed since the 1950s.

→ Building phase of the Kebun-Kebun Bangsar community garden in Bangsar, a neighbourhood in Kuala Lumpur, 2017

When you returned after your studies in Aotearoa New Zealand in 1994, you were one of the first local landscape architects practising in Kuala Lumpur. What kind of garden and landscape design culture did you encounter on your return?

Compared to some of our neighbours, like Bali, or even Japan and China, Malaysia does not have a traditional history of garden design. It was a rural and agricultural landscape. The big change in this part of the world has happened in my lifetime. In the past 60 years, when cities like Kuala Lumpur rapidly expanded, landscape architecture became a necessary practice.

But in the 1990s, the big public projects were still done by landscape architects from Europe and America. Now, we have a lot of schools that are training young people to be landscape architects and designers. It is also now a requirement that new developments have green space or gardens.

That wasn't the case before?

No. In Europe, landscape and garden planning underwent a big shift 200 years ago, with industrialization and the urbanization that came alongside it. When cities became barely liveable because of pollution and population density, planning frameworks were developed to provide access to green space. Because of Malaysia's much later yet incredibly fast urbanization, there were no planning frameworks for gardens and landscapes until the 1990s. When I first started working here, landscape was not a requirement of any architectural development. Architects needed to submit building plans to local authorities for approval, but there was no such thing as a landscape submission. Only in the last 20 or 30 years has the legal framework slowly developed. So anyone developing buildings now must also put in a landscape or garden submission.

I heard you only work on local projects, is this correct?

Strictly speaking, I am retired. In 2013, I realized that I'd been doing a lot of projects for private developers, essentially serving the top 5 to 10 per cent of our society. I gave up my commercial practice and have been trying to bring design to the lower end of the spectrum of our society ever since. I am involved in community health projects and making design very simple, accessible, and affordable.

↖ Hilltop view from Kebun-Kebun Bangsar, looking south-east from the street Lorong Bukit Pantai 4, Bangsar, 2021

↗→ Building phase, 2017

We were teaching people how to plan for an uncertain future by growing food.

In 2013, you and a group of people started a campaign called Malaysian Spring. What was that about?

To support a young woman candidate, Nurul Izzah, in the 2013 general election, I joined a group of volunteers to organize a visual campaign by planting colourful "flowers" made from cloth in the city's public spaces. It was both a political project and a public education project. By "planting" these flowers, we also showed the potential of unused public spaces like roundabouts or traffic islands. After the campaign, we had mobilized so many volunteers that we felt we should continue with the movement of trying to give people more hope, but also greater access to gardens or a farm. So we went ahead and applied to the local authorities and to the government for pieces of land to grow our own vegetables and flower gardens.

I read that a primary objective of starting Kebun-Kebun Bangsar was to feed underprivileged people.

Yes, there were still a lot of homeless people and people living in slums in Kuala Lumpur. But we also foresaw that in the future we might encounter food insecurity caused by climate change or environmental degradation. So we were teaching people how to plan for an uncertain future by growing food. Then, in 2020, when the pandemic came, the situation we had been preparing for almost immediately happened in front of our eyes. During lockdown, food was unavailable in the supermarket. People were panicking. But we had food in the garden!

Your garden is a public space. It serves educational purposes, but it is also a place for leisure. The boundaries between park and allotment, agriculture and community garden blur. As someone who has been in the landscape architecture field for a long time, do you feel that this is the urban green space of the future?

I will answer this in the context of Kuala Lumpur. There is a WHO standard for the number of square metres per person in the city. Kuala Lumpur is way behind European cities, as well as our neighbours like Singapore. In Kuala Lumpur, we are running out of land. That is why I and some like-minded people are looking at spaces reserved for trade or utilities

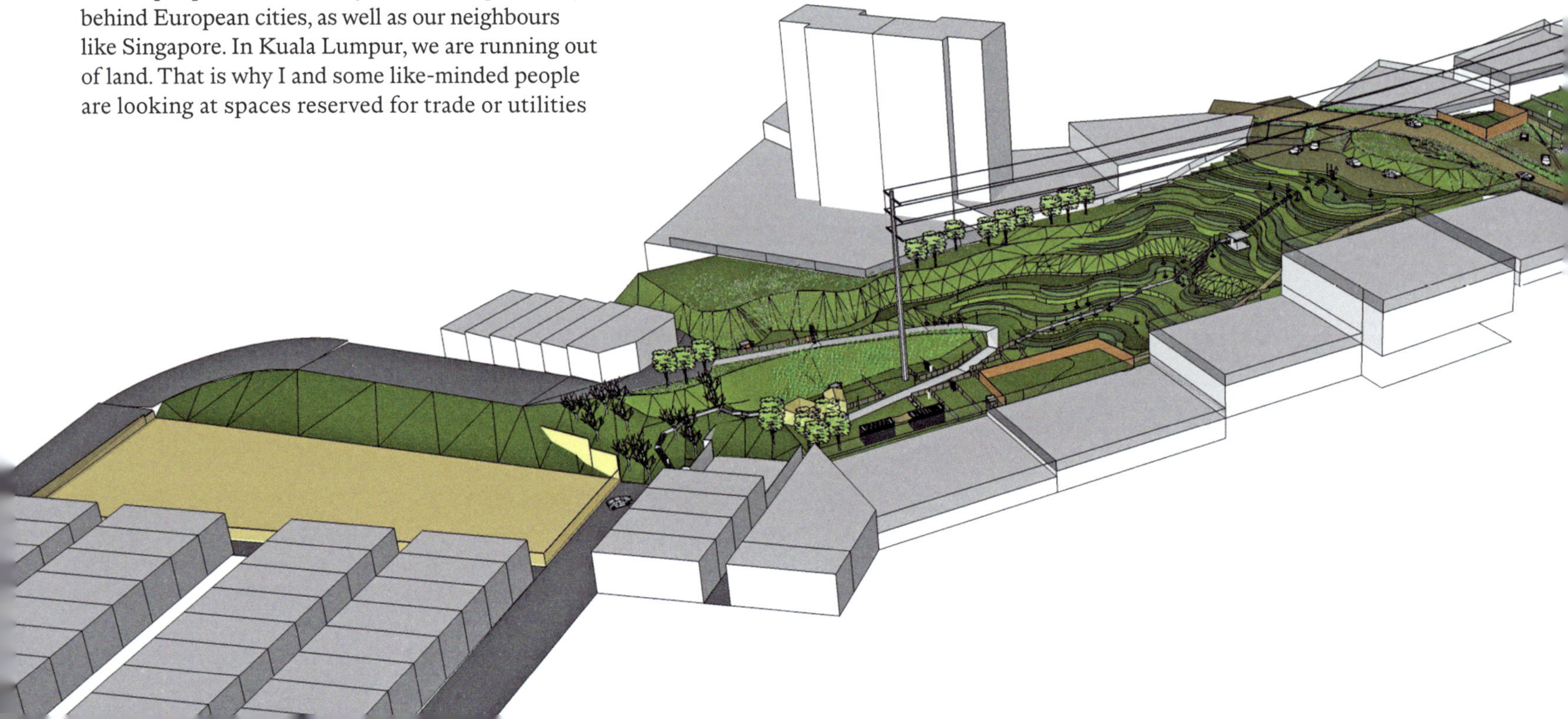

↙ Kebun-Kebun Bangsar
residents, 2020

↓ 3D model of the
Kebun-Kebun Bangsar
site, incorporating the
high-voltage overhead
power line

Too much landscape design
has happened from the top
down, at least in the 30 years
that I have been doing it.

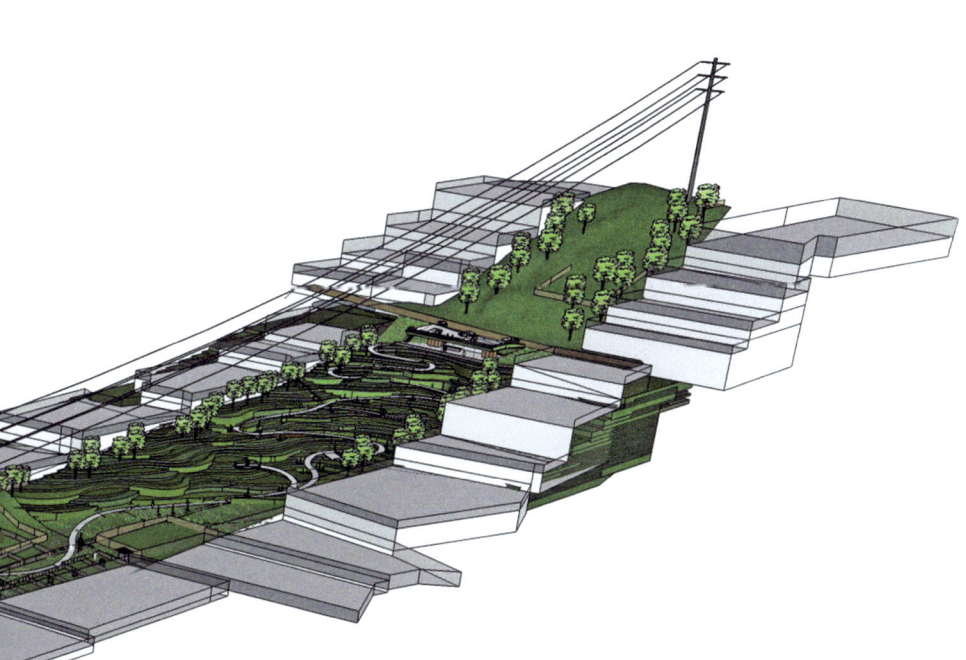

usage to demonstrate to the public, but also to the
authorities that these reserves need to have multiple
uses. Even cemeteries can serve the community in
more than one way. So, with our future projects we
want to counteract the fact that thousands of hectares
of land are underutilized. Kebun-Kebun Bangsar
was a first disruption to the existing system. It is
an example of how some of this space can be used, and
hopefully that movement or that precedent can be
replicated and scaled up by other people.

So, what you are talking about is designing systems, providing an example or a manual for people to copy and adapt to a new local context.

Yes, exactly. It is about democratizing the way
public space is designed. The use of gardens and parks
is administered by local authorities, sometimes with
a lot of restrictions. They often provide parks which
are not useful for the local community. We felt that the
local community must take more action, or take part
in the garden creation or the park creation because they
know what's best for their own little community.
Too much landscape design has happened from the top
down, at least in the 30 years that I have been doing it.

You have just handed over the reins of the garden to a group of women. What is next on your agenda?

Quite a few projects. All of them involve trying to
take resources out of the government's hands and
hand it back to the public. So we are working on a big
area of the cemetery in the middle of Kuala Lumpur.
It is very large and financially unsustainable. We are
working on the government to convince them to turn
it into a park.

Another project is in the nearby University of
Malaya. There's a huge botanical garden there which
nobody knows about. It was started by the university
as a research garden. The university's funds for maintaining the garden are dwindling, but we are telling
them they don't need government grants to run it.
We can use community effort and community funding.

All of our work is about connecting. What is
important is not just the human usage. We're trying
to introduce the idea that our cities are not just about
human beings, but also about other species – about
the insects, birds, or even the larger animals. In the
future, we hope that our national parks are not
only on the outskirts but also within the city. Bringing
nature back into the city. Designing with nature ●

The World as a Garden

Images of the garden as a place of healing, spirituality, or learning are often associated with the gardens of the past, which for many people were a kind of sanctuary in which to create a domesticated version of untamed nature. In the Anthropocene, however, we are bound to ask who needs protecting from whom?

Given that there is not an inch of the planet that has not been impinged on or affected by human activity, the question that really exercises the minds of designers, planners, and researchers is this: if we cannot avoid having an influence, might not that influence be made less destructive, or even turned into a positive? There are certainly examples to prove that this is possible. After all, these days there is more biodiversity in an ordinary suburban front garden than on a field leached of all life by industrial farming. Might we not learn to view the whole Earth as such a garden that we have a duty to cultivate, tend, and use responsibly?

For many artists, landscape architects, designers, and scholars, the garden is an opportunity. It is a place in which to ponder and try out in miniature what might be practised on a much larger scale. It is above all in the garden, where the gardener's attentive, caring hand is constantly at work, that we can counter the challenges of climate change, the threats to biodiversity, growing social injustice, and ever greater isolation with confidence, and so enable new forms of togetherness and a better, more harmonious coexistence with nature to thrive.

GILLES CLÉMENT is a landscape architect, theorist, and professor, but, above all, he is a gardener. In this essay, he proposes a new understanding of gardening – one that sees the entire planet as a garden, and views humans not as the rulers of plant life but as actors within a complex system of living beings.

The Planetary Garden

The "planetary garden" is a means of living in harmony with nature, appreciating the ecosystem in all its diversity, acting as gardener and guardian. I am first and foremost a gardener: on the one hand, because I have a garden, and on the other, because I think that the garden is at the forefront of our current understanding of the terrain as a whole, and consequently of the landscape which creeps into the garden. I am a gardener in the literal sense of the word when I handle the soil, and again, at a certain remove, when I attempt to work on landscapes that are large-scale. Both tasks ultimately revolve around living organisms; that is a strong point they have in common which cannot be sidestepped, a question that cannot be avoided.

Long before humans were born, animal and plant species were travelling ceaselessly: seeds transported by the wind, by animals, and by sea currents. We do not know the geographical origin of the largest seed in the world, the coconut, travelling on the planet's tropical belt, washed away by cyclones. While the coconut floats, maple samaras and dandelion seeds fly away. Many seeds pass through the digestive tract of birds, rejected as soon as the fruit pulp has been digested – like cherries. It turns out that many seeds

benefit from this distress to be able to germinate. Birds travel, and the planet is a territory where multiple species are constantly creating "emerging ecosystems", expressing the mechanisms of coevolution.

Gardens created by humans, although generally modest in size, contribute greatly to the spread of species around the world. From its origin, the garden has been a site of accelerated formulation. We set up a garden in the heart of a clearing, we close it to protect it from predators, and within this enclosure we place the "treasure". The species brought into the enclosure first came from nearby territories, but now the potatoes and tomatoes in our vegetable gardens in Europe come from South America.

Planetary intermingling has increased enormously over the past decades, because human activities mean that we are constantly moving, and making everything else move. Consequently, plants and animals meet in new and unforeseen circumstances, not permitted spontaneously by geography. Butterflies, wind, seeds, people: everything communicates. The horizon is no longer the limit of our landscape; we now know what is happening on the other side. If you pour a cup of bleach down the sink, it goes into the ocean, and the life of the ocean is impacted by it.

→ Gilles Clément in his garden in La Vallée, France, 2021

Gardens created by humans, although generally modest in size, contribute greatly to the spread of species around the world.

Nature is not at the service of man: we exist within her, submerged in her, intimately associated with her. The planetary garden is a means of considering ecology as the integration of humanity – of the gardeners. Its guiding philosophy is based on the principle of the "garden in motion": do the most *for*, the minimum *against*. The ultimate goal of the planetary garden is to exploit diversity without destroying it, perpetuating the "planetary machine", and ensure the existence of the garden, and hence, the gardener.

Garden books don't mention wild creatures, except to explain how to restrict them. Tradition excluded from the gardened area all those living species, both animal and vegetable, which escaped the gardener's control. Publications remain obstinately silent on the moles of Babylon, the beetles of Villandry, the dragonflies of Versailles, or the snakes of the Alhambra. They must still be living there, yet none of them has any bearing on the artifice for which the gardens are famous.

My project at La Vallée, my own home, was not to build a house with a garden around it. It was the opposite: I wanted to live in a garden. But with no example around to help me, I had to experiment. I decided that one part of the garden, being biomass (leaves, fruit, rhizomes, seeds, etc.) would be given back to the animals accustomed to eating it. That meant giving up a portion of space.

Ecology itself destroys the notion of the "enclosed" garden. Birds, ants, and mushrooms recognize no boundaries between territory that is policed and space that is wild. Ecology's primary concern is nature in its entirety, and not the garden in particular. The enclosure was always an illusion; a garden is bound to be a planetary index. For me, success in the work of the architect lies not in the arrangement of forms, or in the balance between shade and light – I am no judge of that – but in the acknowledgement of a life saved: the life of a roebuck, a coypu, a beech marten; the life of a bat, a lizard, a hedgehog; the life of any one of the silent, slow, and sometimes imperceptible insects not seen by our predatory eyes.

It is no longer possible to garden without being aware of the ecosystem. Many think they are gardening when actually they're impoverishing the flora of their environment, just to keep a few beautiful rose bushes or a few hydrangeas, the image of a garden appropriate for the individual. There are also corporations taking pieces of the Amazon and leaving nothing behind but barren land. That is no longer gardening, it is devastation.

While in the human garden the enclosure creates the limits of a space protected from predators, the enclosures of the planetary garden are the natural and undoubtedly fluctuating limits of the biosphere. Beyond these limits (approximately 11,000 metres above sea level and 8000 below) no living system is known. We all inhabit the same enclosure, we could not live anywhere else.

Humans are everywhere on the planet just like the gardener is everywhere in his garden. Even where we do not set foot, we know what is going on. Drones, satellites, all the means of remote observation provide images and information. Spaces that may have been abandoned, no longer visited by the gardener, remain places both known and observed, under surveillance. Diversity also lies in the places that are neglected: the sides of the road, abandoned plots, wasteland, heath, peat bogs, anywhere where it is difficult to exploit the land with machines. It is here that biological and genetic resistance can be found.

If the planet is a garden, we are all gardeners – perhaps we are not aware of it, yet our choices and lifestyles have an impact on the biosphere and on our collective vital space. The gardener discovers something new in their garden every day. They are in constant dialogue with all of its diversity. Unlike a surface technician who arrives on the ground with machines, executes a task, and moves on to another site, the gardener respects diversity and gives it the freedom to exist. They do not seek to keep things "clean", but to protect life while using their power to assist health and human balance. For the gardener, there are no bad weeds – only weeds that are sometimes badly placed.

Humans have sown ideological fields on the planet, created religions, developed a cultural landscape of great diversity. Each of the human cultures is expressed by a world vision which has an immediate impact on the way of life, the habitat, and the landscape. Rice is not harvested with the same techniques in Louisiana as in Bali, and this difference is not only due to variations in topography between these two regions. In Louisiana, rice is associated with industrial production, while in Bali it is linked to Dewi Sri, the goddess of rice, to whom it is advisable to make offerings at the time of fruit setting, while keeping away the *burung padi*, birds, which feed on the grains as soon as they ripen. The way we see the world has an impact on the way we look after it.

Diversity is represented not only by the different species, but also by the variety of human behaviour

↑ Gilles Clément's
garden in La Vallée, 2021

in the face of this biological diversity. From this point of view, we can say that the planetary garden has nothing to do with globalization. Although the scale of impact is the same, the planetary garden protects and develops diversity in all its forms, while globalization erases it in the name of market forces. The standardization of urban and rural landscapes is a result of the power of industry lobbies around the globe – why do we live in the same type of tower built in San Francisco as in Shanghai?

The conservation of ancestral practices is important to planetary gardeners. We do not grow the same varieties of carrots in the north as in the south of Europe. Standardizing with a single variety for reasons of profitability means eliminating varietal wealth, while suppressing local activity from a non-spending economy. Cultivating a species adapted to the soil and to the climate of the place means eliminating unnecessary and expensive inputs. It is this model that we must support.

Happily, there are experiments proving that human intervention isn't completely irreversible, as long as species have not disappeared. When a piece of land in a very traumatized environment is left to cleanse itself, it begins by becoming impoverished because it was loaded with fertilizer, but then, progressively, it recovers a series of flora corresponding to its natural capacity to accommodate diversity – it takes seven to ten years for the equilibrium to be restored in agricultural soil – and species no longer visible begin to reappear.

As planetary gardeners, we depend on all the biological diversity that we harness for a living. We must preserve it to ensure a future. What kind of model do we want to develop to allow both the exploitation of diversity for the benefit of an expanding world population and the preservation, in terms of quality and quantity, of that diversity on which we depend? We must respond to this question as quickly as possible if we are not to succumb to the inevitability of irreversible forces of destruction, but rather to develop in every conceivable field a mental territory of optimism – a garden ●

Article originally published in *The Architectural Review* (16 February 2021), online: https://www.architectural-review.com/essays/in-practice/in-practice-gilles-clement-on-the-planetary-garden. Courtesy of *The Architectural Review* and Gilles Clément.

CHRISTOPH MILER views existing botanical classification systems to be the result of design. He poses the question: if we reframe conventional – perhaps outdated – perspectives on the plant world, could this help us to reconsider our relationship with nature?

Taming the Taxonomy

The kingdom of plants is a place of overwhelming diversity. Leaves, flowers, fruits, and stems come in a kaleidoscopic expression of shapes, textures, and colours. Some plant species remind the innocent human eye of peach-coloured tongues (*Cryptostylis subulate*), ultra-long phalluses (*Titan arum*), or life-sized pistachio ice cream scoops (*Azorella compacta*). With these endless variations, botany – the scientific and orderly study of plants – has always posed a challenging endeavour to the human mind.

The first attempts to design a system of plant classification were made in India and Greece over 2300 years ago. In Europe, however, many of the early botanical insights fell into oblivion during the Middle Ages, when scholars focused on the medical aspects of plants and not their taxonomic traits.[1] It was only with the rise of botanical gardens as a place of scientific curiosity that serious efforts of tackling plant systemization were reintroduced in Renaissance Europe. In the gardens of medical universities and monasteries, including Orto botanico di Pisa (opened 1544) and Orto botanico di Firenze (opened 1545), botanists such as Italian Luca Ghini cultivated and classified plants. Their work led to a new type of publication, the herbarium, in which they documented and compared dried specimens in a systematic manner.

Nonetheless, there was no universal system in place; plants were classified inconsistently and names for new species were a matter of taste, often voluminous and complicated. The tomato, for example, was called *Solanum caule inermi herbaceo, foliis pinnatis incises*, which translates in English as "the solanum with the smooth stem which is herbaceous and has incised pinnate leaves". This kind of tongue-twister must have been rather impractical for a chat between two botanists.

In the eighteenth century, now legendary Swedish botanist Carl Linnaeus set out to clean up what he perceived as a mess. While working as head gardener for the director of the Dutch East India Company, George Clifford, at his summer home in Hartekamp near Haarlem, Linnaeus had access to a vast botanical library, greenhouses, a grand garden, and the company's vegetal prey from overseas explorations. His studies of the sumptuous flora would provide the basis for a radical idea: the kingdom of plants could be neatly organized by applying "the social hierarchy of his day, with its kingdoms, provinces, parishes and villages, to the natural world. He slotted plants [...] into a framework of five main categories – kingdom, class, order, genus, species"[2] – and from there each plant was called by just two succinct Latin names,

→ Carl Linnaeus established three kingdoms in his *Systema Naturae*, namely *Regnum animale*, *Regnum vegetabile*, and *Regnum lapideum*. Tabular representation of the plant kingdom, from Linnaeus, *Systema Naturae*, 1735.

REGNUM VEGETABILE.

N.	O.	P.	Q.	T.	U.	V.	Y.
POLYANDRIA.	DIDYNAMIA.	TETRADYNAM.	MONADELPH.	SYNGENESIA.	GYNANDRIA.	MONOECIA.	POLYGAMIA.
St. multa recept. adnata.	*Stam. 4, quor. 2 longiora.*	*Stam. 6, quor. 4 longiora.*	*St. Filam. coal. in 1 corp.*	*St. Antheræ coalitæ.*	*Stamina Piſtillo adnata.*	*Plantæ Androgynæ.*	*Species Hybridæ.*

N. POLYANDRIA — *St. multa recept. adnata.*

MONOGYNIA.
α. CALICE CADUCO.
Actæa. *Chriſtophoriana* T.
Podophyllum. *Anapodophyl* T.
Corchorus.
Sanguinaria D. *.
Chelidonium T.
Glaucium T.
Papaver T.
Argemone T.
Sarracena T. *Coilophyllum* Mſ.
Tilia.

β. CALICE PERSISTENTE.
Peganum. *Harmala.*
Nymphæa.
Leuconymphæa B.
Michelia †. *Samſtravadi* HM.
Anacampſeros. *Telephaſte.* D.
Ciſtus.
Helianthemum T.
Caryophyllus. *Car. arom.* T.
Thea *. *Belutta* HM.
Capparis.
Plinia. Pl. *

γ. CALICE TABESCENTE.
Euphorbium L. 3.
Cereus.
Opuntia T. *Tuna* D.
Cactus. *Melocactus* T.
§. N. 3. *Delphinium.*

DIGYNIA.
Pæonia.
Anona. *Guanabanus* Pl.

TRIGYNIA.
Pereskia Pl. *
Reseda T.
Luteola T.
Hypericum N. 5.
Androſæmum T.
Aconitum N. 5.
Delphinium N. 1.
Staphiſagria Rp.

TETRAGYNIA. *Tetragonocarpos.*

PENTAGYNIA.
Aquilegia.
Nigella.
Aizoum *. *Ficoidea* N.
Meſembryanthemum D.

§. N. 3. *Hyper. Aſcyrum.*
§. N. 3. *Aconitum.*

HEXAGYNIA.
Stratiotes. *Aloides* B.

POLYGYNIA.
Dillenia †. *Syalita* HM.
Magnolia Pl. *Tulipifera.*
Clematitis.
Atragena. *Viticella* D.
Pulſatilla.
Anemone.
Anemone-ranunculus D.
Nemoroſa Rp.
Caltha Rp. *Populago* T.
Helleborus.
Trollius Rv. *Helleboro-Ran.* B
Helleboroides B. *Aconit.* Rv.
Ranunculus.
Ficaria D.
Ranunculoides V.
Ranunculo-aſphodel. HS.
Adonis D.
Hepatica D.
Filipendula † .
Ulmaria T.

O. DIDYNAMIA — *Stam. 4, quor. 2 longiora.*

GYMNOSPERMIA. i. e. Seminibus Pericarpio nudis.

α. PETALI LAB. SUP. NULLO.
Bulga. *Bugula* T.
Polium.
Teucrium.
Triſſago. *Chamæpitys* T.

β. PETALI LAB. SUP. ERECTO.
Origanum T.
Majorana T.
Thymus T.
Satureia T.
Serpillum T.
Thymbra T.
Lavendula.
Stœchas.
Hyſſopus.
Clinopodium.
Marrubium.
Betonica.
Glechoma. *Calamintha* T.
Chamæcliena B.
Ruyſchia. *Rui,chiana* B.
Ocymum.

γ. PETALI LAB. SUP. CONCAVO.
Mentha.
Menthaſtrum Rp.
Pulegium Riv.
Moldavica.
Volkamera Hs.
Stachys.
Galeopſis.
Ladanum D. *Tetrahit.* D.
Lamium.
Molucca.
Cardiaca.
Galeobdolon D.
Leonurus.

δ. PETALI LAB. SUP. GALEATO
Dracocephalon.
Scutellaria Rv. *Caſſida* T.
Brunella.
Phlomis.

ANGIOSPERMIA. i. e. Seminibus tectis Pericarpio.

Antirrhinum.
Linaria T.
Elatine Rp.
Aſarina T. *Cymbalar.* Rv.
Scrophularia.
Digitalis.
Gratiola Rv.
Volkameria †. *Digitalis ſp.* T.
Chelone A.
Orobanche.
Squammaria Rv. *Anblatum* T.
Acanthus.
Melampyrum.
Fiſtularia. *Criſta galli* Rv.
Pedicularis.
Euphraſia.
Odontites D.
Verbena.
Sherardia V.
Selago. *Camphorata,*
Bontia *.
Dodartia *.
Phelypæa T *.
Creſcentia *. *Cujete* Pl.
Celſia †.
Limoſella. *Plantaginella* D.
Rhinanthus. *Elephas* T.
Martynia. Houſt. apud Mr.
Æginetia †. *Tſcemcamula* HM.

P. TETRADYNAM. — *Stam. 6, quor. 4 longiora.*

FRUCTU SILICULOSO.

α. PERICARPIO UNILOCULARI.
Iſatis.
Crambe.
Cakile.
Myagrum.
Bunias. *Rapiſtrum* T.

β. PERIC. BILOC. DISSEP. OPPOSITO.
Thlaſpi T.
Burſa paſtoris T.
Iberis D.
Biſcutella. *Tklaſpidium* T.
Naſturtium.
Iberis Rp.
Coronopus H. Rp.
Lepidium.
Armoracia Rp.
Cochlearia.
Subularia Rj. *Juncifolia* Rj.

γ. PERIC. BILOC. DISSEP. PARALL.
Alyſſum.
Draba D.
Lunaria T. *Bulbonac.* Rp.

FRUCTU SILIQUOSO.

Eryſimum.
Irio. *Eruca* T.
Sinapis.
Rapa.
Napus.
Braſſica.
Turritis.
Hesperis.
Alliaria Rp.
Conringia Hs.
Dentaria.
Sophia. *Accipitrina* Rv.
Siſymbrium.
Radicula D.
Cardamine.
Raphanus.
Raphaniſtrum T.
Cleome. *Sinapiſtrum* T.
Cheri. *Leucojum* T.

Q. MONADELPH. — *St. Filam. coal. in 1 corp.*

PENTANDRIA.
Hermannia.
Melochia D. *
Xeræa. *Amaranthoides* T.

DECANDRIA.
Azedarach.
Geranium X: 1.
Gruinalis.

POLYANDRIA.
Malva.
Alcea T.
Abutilon T.
Malope †.
Lavatera. A.
Goſſypium. *Xylon* T.
Alcea. *Malva roſea.*
Althæa †.
Urena *.
Trionum. *Bammia* Rv.
Ibiſcus. *Ketmia* T.
Camellia *. *Tſubaki.* Kp.
Sida. *Altheades.* Mg.
Fevillæa. *Inga* Pl.

R. DIADELPHIA. — *St. Filamentis coalita in 2 corpora.*

HEXANDRIA.
Fumaria T.
Capnoides T.
Splis Rv.
Capnorchis B.
Cyſticapnoi B.

DECANDRIA. i. e. Stamina coalita filamentis 9 in unum corpus, & 1 libero.

α. FRUCTU SILICULOSO.
Polygala.
Cicer.
Lens.
Onobrychis.
Sertula. *Melilotus* T.
Dorycnium. *ζ* ? ſimul.
Trifolium. *ζ* ?
Coreba. *Lagopus* Rv.
Anthyllis Rv. *Vulneraria* T.

β. FR. INCURVO IRREGULARI.
Medica T. *Falcata* Rv.
Medicago T. *Cochleata* Rv.
Hippocrepis. *Ferrum Equ.* T.
Scorpiurus. *Scorpioides* T.
Ornithopodium.
Telis. *Fœnum Græcum* T.
Hedyſarum.
Melihonia Hs.

γ. FR. LEGUMINOSO ORDIN.
Lotus.
Ononis.
Ternatea. *Clitoris,*
Corallodendron.
Colutea.
Ulex. *Geniſta-Spartium* T.
Spartium.
Geniſta.
Anagyris.
Cytiſus.
Laburnum.
Orobus.
Vicia.
Arachis. *Cracca* Rv.
Lathyrus.
Clymenum.
Niſſolia.
Lupinus.
Faba.
Piſum.
Phaſeolus.

δ. FR. BILOCULARI.
Biſerrula †. *Pelecinus* T.
Tragacantha.
Glycia. *Aſtragalus* T.

S. POLYADELPH. — *Fil. coal. in plures part.*

POLYAN.
Laſianthus †. G. *Alcea ſp. aliis.*
§. XX 1. *Citrus.*

T. SYNGENESIA — *St. Antheræ coalitæ.*

MONOGAMIA.
α. FLORE SIMPLICI.

β. SEMIFLOSCULOSI T.
Lampſana.
Cichorium.
Catanance.
Zacintha.
Taraxacum. *Dens Leonis* T.
Piloſella.
Hieracium.
Sonchus.
Chondrilla.
Picris †.
Lactuca.
Scorzonera.
Tragopogon.

γ. FLOSCULOSI T.
Chryſocome. *Linoſyris* Mg.
Eupatorium.
Sphærocephalus. *Echinoſus* T.
Santolina.
Vebeſina. *Bidens* T. Pn.
Forbicina Pn.
Carlina.
Xeranthemum T. *Stœbe* Rv.
Serratula D.
Carthamus.
Carduus.
Cinara.
Arctium. *Lappa* T.
Cnicus.
Petaſites.
Klenia †. *An Tithymaloides* B.

α. RADIO PETAL. DESTITUTO.
Artemiſia.
Abſinthium.
Abrotanum.
Filago.
Ananthocyclus V.
Tanacetum T.
Baccharis D.
Senecio.

β. RADIATI T. CALICE SEMIGLOBOSO.
Achillea. *Millefolium* T.
Ptarmica T.
Anthemis. *Chamæmelum* T.
Buphthalmum.
Matricaria.
Bellis.
Leucanthemum.
Chryſanthemum.
Cotula.

γ. RADIATI T. CALICE VENTRICOSO.
Calendula. *Caltha* T.
Dimorphotheca V.
Tuſſilago.
Doronicum.
Arnica Rp.
Solidago. *Doria* D.
Virga aurea T.
Jacobæa.
After.
Amellus †.
Helenium. *Enula Camp.* Mg.
Erigerum. *Conyzoides* D.
Othonna. *Tagetes* T.

α. RADIATI T.
Helianthus. *Corona Solis* T.
Rudbeckia. *Obeliſcotheca* V.

β. FLOSCULOSI T.
Jacea.
Cyanus.
Centaurium. *Cent. maj.* T.
Crupina D.

Parthenium. *Partheniaſtrum* D,
Milleria. Houſt. apud Mr.

U. GYNANDRIA — *Stamina Piſtillo adnata.*

DIANDRIA.
Orchis.
Satyrium Rv.
Satyrium.
Orchioides. Trew.
Neottia. *Corallorhiza* Rp.
Serapias. *Helleborine* T.
Herminium. *Monorchis* M.
Cypripedium. *Calceolus* Mar.
Epidendron †. *G. Orchidis eſſ.* Hr.
Ophis.
† *Nidus Avis* T.

TRIANDRIA.
Bermudiana.

TETRANDRIA.
Nepenthes †.

PENTANDRIA.
Aſclepias. *Vincetoxic.* Rp.
Beidelſar Kn.
Periploca.
Stiſſeria. *Craſſa* Rv.
Paſſiflora. *Granadilla* T.
Murucuja, T.
Clutia B.

HEXANDRIA.
Ariſtolochia.

DECANDRIA.
Heliēteres. Plk. *Iſora* Pl.

POLYANDRIA.
Grewia †. † *Guidonia* B.
Arum T.
Dracunculus T.
Colocaſia Rj.
Ariſarum T.
§. V: 1. *Rham. Cervi Spina* D.
Calla. *Anguina* Trew.
Arioides B.
Acorus VI: 1.
Ruppia *. *Bucca ferrea* M.

V. MONOECIA — *Plantæ Androgynæ.*

MONAND. TRIAND. IVAND. VAND. POLYAND. Staminaplusquam 7.
MONAD. POL. MONOEC.
Zannichella M *. *Aponoget.* Pn.
Najas *. *Fluvialis* V.
Cynomorion M. *

Thalyſia. *Mays* T.
Sphærium *. *Lacryma Jobi* T.
Ægilops. S. *
Iſchæmum *. *Dactyloides.*
Carex. *Cyperoides* T.
Scirpoides Mg. *Carex* Rp.
Diaſperus. *Niruri* Mr.
Alnus.
Betula.
Buxus.
§. V. 5. *Urtica.*
Amaranthus.
Jatropha *. *Manihot* T.
Andrachne. *Telephoides.*
Oxydectes. *Ricinoides* T.
Ceratophyllum. *Dichotoph.* D.
Myriophyllum Pn.
Pentapterophyllum D.
Corylus.
Oſtrya M.
Carpinus T. M.
Fagus.
Caſtanea T.
Quercus.
Ilex T.
Suber T.
Sagittaria Rp. D.
Sparganium.
Typha.

Pinus.
Abies.
Larix. *
Thuya. *
Cedrus. *
Xanthium.
Ricinus.
Bryonia.
Momordica.
Sicyos. *Sicyoides* T.
Tamnus.
Luffa Arab.
Anguria.
Colocynthis.
Melo.
Pepo.
Cucurbita.
Anguina M.

X. DIOECIA. — *Pl. Mares & Feminæ.*

MONAND. IIIAND. IVAND. VAND. VIAND. VIIIAND. IXAND. XAND. POL. MONAD. SYNG.
Salix.
Phœnix *. *Palma.*
Oſytis. *Caſia* T.
Morus.
Hippophaë *. *Rhamnoides* T.
Myrica. *Gale* M.
Urtica. V. 4.
§. V: 1. *Rham. Cervi Spina* D.
Lentiſcus.
Toxicodendron.
Humulus. *Lupulus* T.
Cannabis.
Spinacia.
Smilax.
§. VI: 3. *Rum. Acetoſa.*
Populus.
Laurus.
Mercurialis.
Hydrocharis. *Morſus ranæ* D.
Saſſafras †.
Nyſſa †. G. *Tupelo* Catb.
§. X: 3. *Cucub. Lychnis.*
§. X. 5. *Cann. Cannabina* T.
Papaya T. *
Aruncus. *Barba Capra* T.
Kiggelaria †. *Arb. Jlicis folio* B.
Juniperus T.
Sabina Rp.
Taxus *.
Ruſcus *.

Y. POLYGAMIA — *Species Hybridæ.*

	VI: 3
Veratrum.	
Valantia A. *.	IV: 1
Holcus †.	III: 2
Sorgum M.	III: 2
Schœnanthum M.	III: 2
Halimus * Mg.	X: 2
Atriplex.	V: 2
Parietaria.	IV: 1

§. IV: 1. *Poteriûm* N: 2

DIOECIA.
Fraxinus T. Pn.   II: 1
Ornus Pn.   II: 1
Elichryſum.   T: M.
§. X. 5. *Sedum, Rhodia.*

TRIOECIA.
Empetrum.   III: 1

Z. CRYPTOGAMIA — *Flores abſconditi.*

ARBORES.
Ficus.
Capriſicus Pn.   III: 1
Erinoſyce Pn.

FILICES.
Equiſetum.
Ophiogloſſum T.
Lunaria Rp.
Pteris. *Thilypteris* D.
Polypodium.
Lonchitis.
Hemionitis T.
Lingua Cervina T.
Adiantum.
Trichomanes.
Acroſtichum †.
Muraria.

MUSCI.
Lycopodium D.
Selaginoides Rj.
Selago D.
Lycopodioides Rj.
Fontinalis D.
Sphagnum D.
Mnium D. *Muſcoides* V,
Hypnum D.
Bryum D.
Polytrichum D.
Jungermannia. *Hepatica* M.
Lichenaſtrum D.
Marchantia. *Lichen* D.
Marſilea. *Lunularia* M.
Lichen. *Lichenoides* D.

ALGÆ.
Fucus.
Ulva Rj.
Hydrophace Bx.
Lemna. *Lenticula* M.
Lenticularia M.
Chara Rj. *Hippuris* D.
Conferva Rj.

FUNGI.
Agaricus D.
Amanita B.
Boletus D.
Hydna. *Erinaceus* D.
Merulius B. *Morchella* D.
Elvela. *Fungoides* D.
Peziza D. *Cyathoides* M.
Coniplea. *Lycoperdon* M.
Lycoperdaſtrum M.
Geaſter M.
Carpobolus M.
Byſſus Rj.
Noſtoc V.

LITHOPHYTA.
Spongia.
Badiaga Bx.
Iſis. *Keratophyton* B.
Tubipora. *Tubularia* T.
Cellepora †.
Millepora.
Madrepora.
Retipora.
Corallium.
Acetabulum. *.
Eſchara.

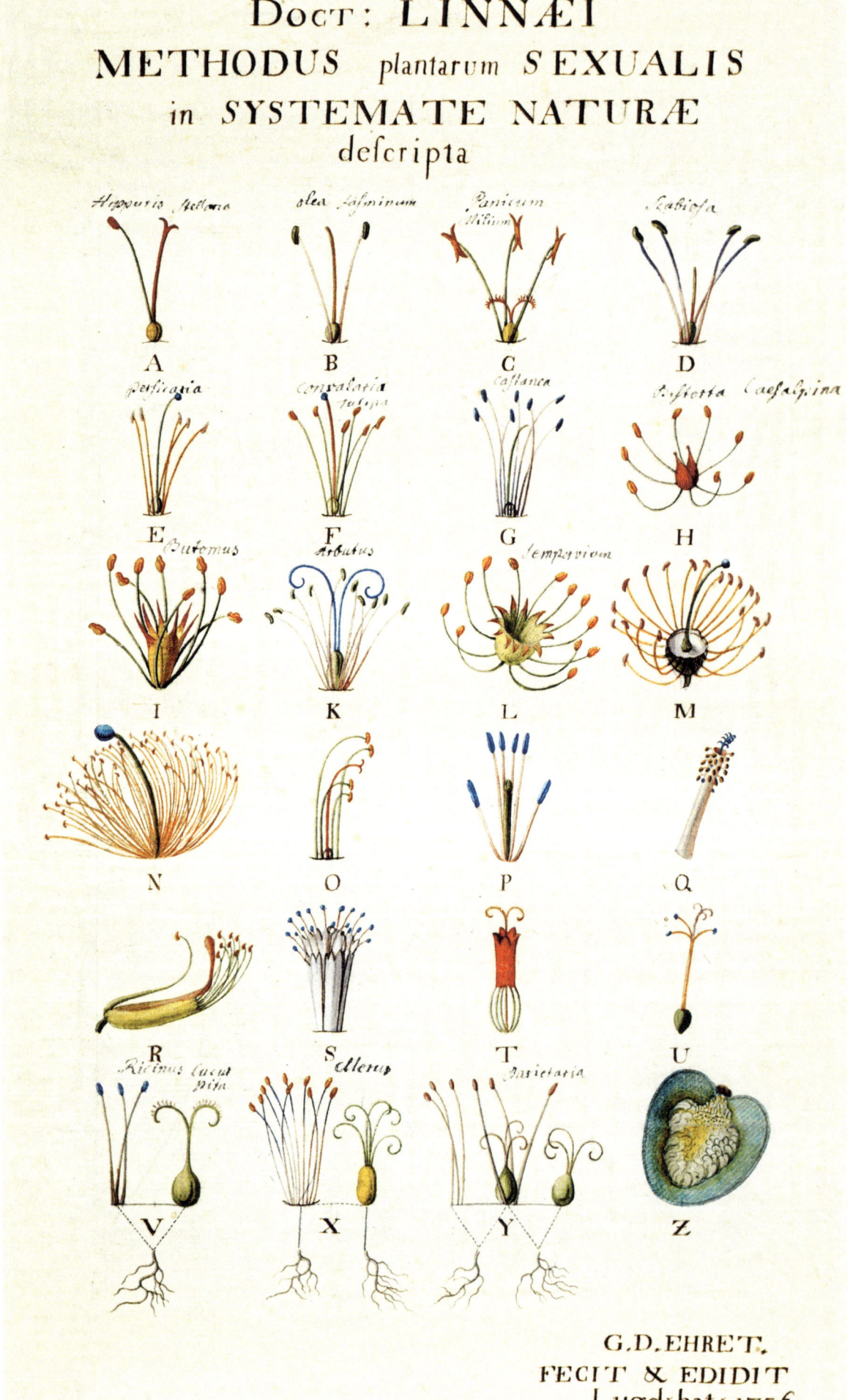

← Illustration by Georg Dionysius Ehret, depicting Linnaeus's sexual classification of plants from *Systema Naturae*, 1735

↗ "A Garden of Parallel Plants" from Leo Lionni, *La botanica parallela*, 1976

representing genus and species. He then went on to consider the number and arrangements of sexual organs as the most important character for designating the class of a plant within his system. In a rather sexualized and patriarchal version of nature, Linnaeus formulated an index based on the number of "husbands" (male stamens) and *concubinae* (female pistils) that were sharing the "bed" within a "house".

The method of Linnaeus to bring order into the overflowing world of living things wasn't merely an act of scientific need, but should also be viewed as an act of design. If we consider the roots of design in the Latin term *designare*, which translates as "to mark out, to choose, to designate", then we might also see the parallels to his classification system. Linnaeus's work equally *designates* and *chooses* "constant, certain and organic"[3] plant features as representative, and thereby predefines what one should actively look at and what one should simply ignore. In order to render the sweeping kingdom of plants into numerical and standardized terms, one could argue that Linnaeus had to *design* this highly selective, new botanical gaze.

Shortly after Linnaeus published his theories in *Species Plantarum* in 1753, he was critically acclaimed by his contemporaries, amongst them philosopher Jean-Jacques Rousseau who praised him as the "greatest man on earth".[4] Linnaeus claimed success in what so many before had failed to do: taming the vast plant kingdom through the human mind. The foundations for modern taxonomy had been laid, fitting to the emerging world view of the time – rationalism.

Famous rationalist René Descartes wrote in 1637 that if we can describe, understand, and organize life, we can also "render ourselves the masters and possessors of nature".[5] The idea was that if we could comprehend and control (and ultimately exploit) our planet through intelligence and technological innovation, we could continue to claim top position in nature's hierarchy. Fast-forward nearly 400 years: if Linnaeus, a firm believer in the idea of plants as God's creation, was to take a walk through our contemporary landscape, he would likely encounter genetically engineered species resistant todisease, plants with new DNA strands that glow at night, plants that grow without soil, and plants that are mere 3D replicas of their long-lost originals. Welcome to the age of the Anthropocene, Mr. Linnaeus, we have taken God's creation into our own hands!

However, cracking glaciers, rising sea levels, and increasing extinction rates draw a very different picture of our position within the world. The manifestations of our ecological crises tell us that the rationalist fantasies of total control and endless growth have reached their limits – our relationship with plants can no longer be a simple one-way street of domination. We need to change something. But where to start?

Perhaps the first thing to recognize is that the rationalist perspective on the world is a powerful one, for it suggests control, order, and hierarchies when in fact there are none. It's a way to look at the world

leo lionni

René Descartes wrote in 1637 that if we can describe, understand, and organize life, we can also "render ourselves the masters and possessors of nature".

that we have found comforting, a story that we've constructed to feel safe as part of a complex and sometimes chaotic reality. By circulating this narrative we've learned to perceive the world in a certain way: mainly through numbers, abstractions, and logical arguments, which created a particular reality for us in turn.

Philosopher Timothy Morton proposes a different perspective, not one of hierarchy but of an ecological mesh that interconnects all living beings, including us. Here, human beings are not an outside force disturbing an otherwise natural system, rather they are an integral and interacting part of the system itself. But if we tell the stories of plants solely through the lens of scientific objectification, static taxonomies made of fixed categories and their related economic value, it is hard to see the dynamic mesh that plants and humans are part of. This is what biologist and philosopher Donna Haraway implies when she says, "It matters what stories tell stories".[6]

Looking for new perspectives and novel stories of plant life is to reconsider what Linnaeus did and to add a new chapter. Instead of taming life's variable expressions, it is time to tame the taxonomy. We don't have to kill the ideas of rationalism and its applications, but we do have to find ways of subverting and expanding its hegemonic, reductionist horizon towards more complex symbiotic relations, which we all rely on in the end.[7]

The potential to narrate the structures and experiences of the mesh seems to be particularly promising for designers. Speculative fiction has already paved the way for this. As a thought experiment it allows us to use our current reality as raw material for building stories about potential ways of coexisting. With the tools of science fiction, speculative evolution, and botanical fiction we can blow up bacteria, shrink humans, and let trees talk. We are also free to ask "what if?" Such a speculative way of thinking enables us to explore possible futures, while reflecting on our current day and age. It becomes a tool for structuring our present experience and making sense of reality from a different vantage point.

Authors, artists, and designers have used botanical fiction since the 1850s to discuss humanity's role within the larger scheme of things. H. G. Wells criticized the colonial exploitation of nature in his short story "The Flowering of the Strange Orchid" (1894), Leo Lionni drew and sculpted fictional plants to question the reductionism of botanical reality in his fantastical herbarium *La botanica parallela* (1976; Eng. *Parallel Botany*, 1977), and Ursula K. Le Guin explored ideas of consciousness in plants in her sci-fi story "Vaster than Empires and More Slow" (1971), in which the forest of World 4470 senses and emanates fear as "one big green thought".[8] Today, artists like Hito Steyerl, Suzanne Treister, or Alexandra Daisy Ginsberg – the latter visualizes plants as the means for the production of consumer goods in her project *Growth Assembly* (2009) – continue in the tradition of botanical fiction.

These approaches go far beyond a mere reflection of modern science and technology, they question what plants essentially are and how they escape their normative categorizations. And all suggest a view of seeing the interconnectedness between the planet and the creatures that live on it.

If we want to make these entanglements graspable, methods of botanical fiction, eco-fiction, and speculative evolution are clearly needed. Used in the right way these strategies can help us reconsider our way of looking at the plant kingdom, and in turn our role within the ecosystem – from domination to collaboration. A redesigned botanical gaze will contribute to narratives that break with the outdated idea of human mastery over nature and a hierarchy of living beings. Obviously, there is much to do within the "basic survival project in our present context"[9] as ecofeminist and philosopher Val Plumwood phrases it, but reconsidering our position within the living world through new narratives might be a healthy and necessary starting point. Not least because this might allow us to experience the world for what it is: a highly dynamic network of infinite connections and interdependencies ●

1 Alan G. Morton, *History of Botanical Science: An Account of the Development of Botany from Ancient Times to the Present Day*. London: Academic Press, 1981, pp. 58–64.

2 Kennedy Warne, "Organization Man", *Smithsonian Magazine* (May 2007), online: https://www.smithsonianmag.com/science-nature/organization-man-151908042, accessed 31 August 2022.

3 Mary Letitia Green (ed.), *The "Critica botanica" of Linnaeus*, trans. Arthur Hort. London: Ray Society, 1938, p. 116. Carl Linnaeus's *Critica botanica* was first published in 1737.

4 "What People Have Said about Linnaeus", *Linné on line*, Uppsala University, online: https://web.archive.org/web/20110513033923/http://www.linnaeus. uu.se/online/life/8_3.html, accessed 27 September 2022.

5 René Descartes, *Discourse on Method and Meditations*, trans. Laurence J. Lafleur. New York: Liberal Arts Press, 1960, p. 45 [French original: *Discours de la méthode pour bien conduire sa raison et chercher la verité dans les sciences*, VI, 2, Leiden, 1637].

6 See Donna J. Haraway, "It Matters What Stories Tell Stories; It Matters Whose Stories Tell Stories", *a/b: Auto/Biography Studies*, vol. 34, no. 3 (2019), pp. 565–75.

7 We should mention, as described by Michael Pollan in his 2013 essay "The Intelligent Plant", that plant science has been on the move over the last 15 years. New fields of research like plant neurobiology and plant communication are being explored and knowledge produced that starts to describe the complex lives and relationships of plants in scientific terms through both logical arguments and experiments. Michael Pollan, "The Intelligent Plant", *The New Yorker* (23 December 2013).

8 Referring directly to Andrew Marvell's poem "The Garden" from 1681 which speaks of a "green thought in a green shade".

9 Val Plumwood, "Nature in the Active Voice", *Australian Humanities Review*, no. 46 (May 2009), pp. 127–28.

← Alexandra Daisy Ginsberg and Sascha Pohlflepp, *Growth Assembly (Herbicide Gourd)*, 2009

In northern Ethiopia, agricultural deforestation has left behind arid plains where forests once thrived. For DR. ALEMAYEHU WASSIE ESHETE, the well-being of the region now hinges on the gardeners caring for the remaining pockets of biodiversity.
Interview by Viviane Stappmanns, Photography by Kieran Dodds

"If we destroy the forest, we risk everything."

For the past three decades, Dr. Alemayehu Wassie Eshete has been working to preserve and restore the once abundant woodlands of his native Ethiopian Highlands. His ally in this mission has been the Orthodox Tewahedo Church. Central to their belief system is the principle that forests are sacred places; the trees surrounding the church must be cared for by the priests. When the entire forest in the northern province of Ethiopia was logged and nearly eradicated, this stewardship helped save some small pockets of ancient forest. Today, these sacred and multispecies havens are key sites of biodiversity, and the efforts of the priests – treating the sites as paradisial gardens to be preserved – are not only crucial to the survival of the forest but to a sustainable agriculture.

You started working as a forest ecologist in 1992. The tree coverage in the Ethiopian Highlands had been on a steady decline. From around 40 per cent one century ago to less than 5 per cent today. How did this happen?

In Ethiopia, population growth has been at a continuous 3 per cent per year. In the 1960s, the total population was around 50 million people. Now it is 110 million. Most people live off agriculture. They herd cattle and grow crops, various pulses, and cereals. But it's not modern farming; not like in Europe, where the existing land is made more arable. In Ethiopia, we don't intensify farming, we expand. When there are more mouths to feed, you clear more forest and convert more land to farming. And that is what happened.

→ Debre Ensesa, South Gondar, 2018

You quickly realized that scarce areas of forest could be found around churches. Why?

According to the law of the Ethiopian Orthodox Tewahedo Church, the forests around that church, including all creatures living in it, have to be protected. The priests think of these forests as Gardens of Eden. It is a kind of metaphor. They are God's gift. We think God exists there. So, the priests take care of the forest.

But they didn't notice that these forests, too, are degrading. Every year they are shrinking by one or two metres, pushed by neighbouring farmers. Nobody noticed that. Until I and an American colleague of mine, Dr. Margaret Lowman, went through the painstaking work of reconstructing this shrinkage with the help of Google Maps and showed them the changes over the past decade. Although the priests loved the forests, they hadn't noticed. It was our science that helped them to understand that they are degrading slowly.

So you started working with the priests to help them not only conserve but restore, and in some cases even expand those forests. How did you do that?

My job consists mainly of teaching people. For instance, I go out and give talks to the priests, or I work with school children so that they start appreciating and caring for the forest.

I read that there are 20,000 such church forests in Ethiopia. How do you roll out an education programme on this magnitude?

Actually, I don't work with all 20,000 forests. If I could get help from the international community, I could include more. But for now I am focusing on a select number. These forests represent a new approach to agroecology. So far we have reached about 35 to 40. Every time I get more funding, we integrate more forests into the programme. The ambition is to reach all 20,000; not directly but by creating role models that the caretakers of each forest can learn from.

This is interesting. In our exhibition, we have been looking at the design of gardens in different locations and over time. There is something very particular about how gardening knowledge has been passed on. Regardless of time and geography creating gardens appears to be the ultimate open-source design. Manuals and role models serve to guide the users, who implement the work mostly by themselves. On a very practical level, what do you tell the priests?

Firstly, it is about running awareness seminars based on my research. We give tips and instructions for conservation. For this, we have to run refreshers every year, because the knowledge gets lost.

An act of maintenance and care, like gardening itself.

Ethiopia does not really have a culture of gardening. We are crop and livestock people, agrarian people. You don't really see so-called gardening as in Europe.

← Ura Kidane Miheret I,
Lake Tana, Ethiopia, 2018

↙ Fields near the Bahirda
Michael II, West Gojam,
Ethiopia, 2018

↘ Entos Eyesus, Lake Tana,
Ethiopia, 2018

We are creating role models that the caretakers of each forest can learn from.

You also encourage the priests to build walls around the church compounds. Here in Europe, walls around gardens were initially built to protect ourselves from the wild nature beyond. Your situation is similar, but also somehow inverted. You are building walls to protect the ecosystem from man-made activity.

Every year farmers are pushing from the outside. During the dry season, the church forests are the only green area. Livestock will come in and eat up small seedlings. So, we make walls. It's not a barrier; we call them "conservation walls". Making stone walls is very cheap. Whenever you need to, you can expand them. They are also a habitat for snakes and frogs.

You say the expansive method of agriculture in Ethiopia is threatening the forests. But what would be the alternative – concentrating agricultural efforts by using chemicals and thereby degrading the soil? Is it like a choice between two evils?

Yes, it's a dilemma. Sometimes you have to choose the lesser evil. But we think you can achieve agricultural intensification by using ecologically sound, organic methods. So rather than dropping chemical fertilizer or pesticides, there are other means. In Europe, a lot of damage has already been done to the soil. Ethiopia is in an early stage when the damage is not there yet, so there is scope for applying organic methods. But whatever we do, we must not destroy the forest! Because if we do, we risk everything – losing our soil because of erosion and water resources because river springs dry up, and so the groundwater, and so on. And what's more, once the forest is lost, we cannot get it back.

You also treat the forests as seed banks and preservation chambers.

Exactly, for instance, a lot of plants used in traditional medicine grow there. Also, as I told you, people are engaged in cropping, but crops need pollinators. And pollinators always exist in the forest. So, church forests are not only helping the ecosystem but even the farmers themselves by providing pollinators for their crops.

So on top of working with the priests, shouldn't you be working with the farmers?

Actually, the priests themselves are farmers. They are not full-time priests. So when you work with them, you also work with the farmers. They sustain themselves as farmers and serve the church. Also, priests enjoy a lot of respect.

So that they act as multipliers, I understand.

And protecting the forest means working with the entire community. We don't just educate, or raise funds for the walls, we also plant seedlings in the rainy season. In some cases, we also work with the local community so that they reduce pressure on the forest – we give them efficient cooking stoves or solar lanterns so that they don't need as much wood.

Preserving or "gardening" these forests means working with the whole system, right? You're working with school children, with the priests, with the farmers. It's not just about this one forest.

Yes, it is a holistic approach. Anything else wouldn't work. We can't just build the wall ●

Landscape architect CÉLINE BAUMANN designs urban environments, but she also maintains a prolific practice as an artist and educator. Her Basel-based studio is committed to research on plant life and interrelations with humans. Her intersectional lens in turn informs her design work, in which she aims to create dynamic open spaces that respect the ecology of both humans and nature.
Interview by Viviane Stappmanns

"Human habitats need to accommodate other living beings."

Today, many designers are concerned with creating living *environments*, especially our urban environments, in which plants and animals can also thrive. Is this a new conversation?

The discussion is not new but all the more pressing today, as it is now widely acknowledged that human activities are having a catastrophic impact on all the other species. Everywhere we go, especially as westerners with our high levels of consumption, we bring a lot of destruction, corrupting the planet in an unprecedented way. In 2019, the United Nations released a *Global Assessment Report on Biodiversity and Ecosystem Services*, whose conclusions are dramatic. We are on the brink of ecological collapse, with one million species facing extinction within decades.

As you say, this is not new. In the early 1960s, Rachel Carson published *Silent Spring*, a very influential book that drew attention to the damage of herbicides and pesticides on the planet. It is often named as a precursor to the environmental movement that followed in the 1970s and 1980s. *Silent Spring* was published exactly 60 years ago, yet environmental destruction has only increased. Do you think this moment is any different? As a practising landscape architect can you see that there is real change ahead? Or is it just an academic discussion?

Action is being taken – the problem is being tackled from different directions. David Holmgren, the founder of the permaculture movement, describes this very well. In 2008, he published an essay narrating

→ Céline Baumann,
Detail of the installation
Parliament of Plants,
Matadero Madrid.
Centre for Contemporary
Creation, 2019–20

David Holmgren's four scenarios already exist simultaneously in our current reality. The question is which one will ultimately dominate.

↙ Australian permaculture gardener and public intellectual David Holmgren envisaged four future scenarios about energy transitions and converging crises: 1) Brown Tech / Top-down constriction 2) Green Tech / Distributed powerdown 3) Earth steward / Bottom-up rebuild 4) Lifeboats / Civilization triage. Here, these are visualized by illustrator Andrew Merritt (Something & Son), Art direction: EcoLabs / Johanna Boehnert, 2009.

↘ Studio Céline Baumann and architects Davis Manz, Barbara Thüler, and Farquet Architectes, Model of the new Walkeweg School, Basel, winning competition entry, 2022

The project offers playscape amenities for students as well as ecological corridors for an urban flora and fauna.

different scenarios for the future. In all, he assumes that resources will decline and climate change will occur, but he imagines different responses. In one, technology – like geoengineering the weather – plays a central role. Another corresponds more to the ideal of the 1970s: it involves degrowth, consuming less, and an urban exodus; paired with severe climate change, it leads to a neo-feudal system with hamlets, gated communities, and bartering. Anthropologist Anna Lowenhaupt Tsing calls this "life in the capitalist ruins". Another of Holmgren's scenarios – slow decline of resources coupled with mild climate change – offers a more hopeful vision where compact cities use renewable energy and apply an eco-rationalist approach. And yet another tells a cautionary tale of how our future will look if we continue to do "business as usual". Those four scenarios already exist simultaneously in our current reality. The question is which one will ultimately dominate. Personally, I am very critical of the technocentric approach. It is like proposing to fix the cause of the problem with the root of the problem. As feminist Audre Lorde stated in 1979, "The master's tools will never dismantle the master's house."

So on a very practical level, what do you see happening that is hopeful?

We might look to very simple, individual gestures as well as custom-made solutions for each problem, including reducing grey energy and CO_2 emissions, building less, applying the principles of the sponge city at a wider scale. Those measures are unfortunately unspectacular and hardly reach newspaper headlines.

But don't the simple gestures, like green roofs, also require a lot of technology to work?

In fact, green roofs need very little technology, just slightly more investment to ensure the building structure underneath can carry the extra weight. I am part of a team who recently won a competition to design an ecologically sound school complex in Basel. One of the buildings will have a green roof that is just for plants and animals, not for human use; it will have a decent amount of soil. This will allow for a rich biological life, and it will retain water without too much technology. The fact of making it inaccessible to humans will hopefully allow this roof to be rapidly colonized by the local fauna and act as a sanctuary.

So concepts like the sponge city – in which rainwater is retained and reused on an urban scale – are not technology intensive?

Not necessarily. In our densely populated urban environments with all the sealed surfaces, water cycles have been completely disrupted. The rainwater is often unable to infiltrate the ground to be naturally filtered. And overflowing sewage systems have occasionally resulted in devastating floods. To restore the water cycle we need to replace sealed, paved surfaces with porous ones. This is a completely low-tech solution, which is also cost-efficient. It implies that the space for motorized traffic be reduced, a measure which is often met with stubborn resistance.

I have noticed that in terms of addressing complex, networked planning processes, new skills and networks of people are needed too. Do we need new ways of designing?

Yes. We are currently working with a company specialized in circular material reuse on the school project. They collect components from current demolition

We live in a world which is more and more specialized and it's very easy to lose sight of the bigger picture.

projects and compile catalogues of materials that we can then repurpose. This is an exciting approach, allowing new design possibilities and the creation of a somewhat different aesthetic. Interdisciplinarity is also to me a key aspect of future design practises, including in the field of landscape architecture. We live in a world which is more and more specialized and it's very easy to lose sight of the bigger picture. But the only way we can deal with the challenges of our time is by getting out of our comfort zones and understanding how singular elements are part of a whole chain of actions and consequences.

What would you say are the most forward-thinking developments in landscape architecture right now?

Today, the task of landscape architects is to go beyond the mere role of, say, the beautification of the cityscape with blooming flower beds. Nowadays, with the rapid urbanization we are experiencing, correlated with the increasing loss of natural habitats, there is a growing understanding that natural and urban environments need to overlap. Landscape architects have a significant role to play in this discussion. We are starting to understand that human habitats need to accommodate other living beings. Plants and animals also have the right to be part of the city.

Speaking of rights, besides your work as a landscape architect, you are an artist, producing installations and performative works around plants, and more recently about the rights of plants. Do you think we need to see them differently?

There is a common belief that humans are the most evolved species on the planet. This is in my opinion highly questionable, especially when we consider all the harm we are causing. Working with and investigating plants, I have realized how they have their own kind of intelligence, but also different ways of expressing themselves. They cannot speak as we do, to complain and claim their rights, and have therefore been mostly considered as a commodity, exploited for our own gains, or wiped out if deemed redundant.

But this started changing 50 years ago, in 1972, when lawyer Christopher Stone published an article called "Should Trees Have Standing?" in a law journal.

This has been an influential article for contemporary discussion on the rights of nature. Stone recounts that in the history of constitutional human rights, the rights of underprivileged groups have continually improved in the past centuries. For instance, children, people of colour, Indigenous peoples, women, the queer community, and prisoners, who once had no rights, now – at least in a few places – have gained legal standing. He also argues that plants or areas of nature, too, could be given constitutional rights – one thing comes after the other. And natural entities could and should be next. There are precedents.

Can you name a few?

There are some well-known international examples like the rights given to the Whanganui River in

Aotearoa New Zealand in 2017, or the rights given to nature in Ecuador in 2008.

There, Indigenous communities describe the natural environment as Pachamama. So not only nature but also the way Indigenous communities view it has been protected when the country's constitution was revised to establish Pachamama as a legal entity for the first time in history. It now stipulates the respect for nature's existence, and for the maintenance and regeneration of its life cycles, structures, functions, evolutionary processes, and restoration.

How does giving rights to the environment translate into design? What are the tangible ways of designing with this in mind?

In Basel, a law was passed in the 1980s that protects trees with a circumference of more than 50 cm from being cut down, whether on private or public land. This means that if a construction project is planned on a site with protected trees, the project needs to accommodate them, including their root systems, or, if necessary, replace the trees with other specimens of a similar quality. I hope to see more of these constitutional examples in the future, giving trees and other living beings legal standing. It is a good example of how Stone's legal proposition can be applied, and ultimately filter down into design practise by redefining the power relationship between human enterprise and the usually silent, powerless nature ●

← Céline Baumann, *Parliament of Plants*, for the exhibition *What is Radical Today*, Royal Academy of Arts, London, 2019

All landscapes are interconnected on a shared, global scale, says Belgian landscape architect BAS SMETS. In order to create environments that are liveable while keeping our planet in balance, new microclimates can be introduced with a long-term view to less intervention.
by Lisa Dabscheck

"Let's call it biospheric urbanism."

"Exemplary" and "augmented" landscapes are terms coined by Bas Smets to describe the crux of his approach to creating new urban microclimates. The "exemplary" landscape refers to reading the natural components obscured by the visual clutter that has been introduced into a given site over time. Once these inherent elements have been identified, they can be foregrounded and consolidated. That is the point at which the "augmented landscape" – essentially an enhanced version of the exemplary landscape – emerges. In the southern French city of Arles, Bureau Bas Smets transformed the semi-desert climate of an industrial wasteland into a Mediterranean ecosystem with the introduction of 80,000 plants. An early project in London metamorphosed a sunken domestic courtyard that was bereft of wind and sun into a subtropical forest with the use of residual structural heat and the realization that an incongruous yet ideal microclimate existed there. His winning commission to redesign the public space surrounding the fire-damaged cathedral of Notre-Dame, in what, referencing Victor Hugo, Smets refers to as "the cradle of Paris", utilizes, among other innovations, a thin layer of water on the parvis in front of the cathedral. It will cool down hordes of tourists in high summer with an evaporative cooling that replicates the effect of a warm-weather rain shower.

While Smets's work is grounded in the doctrines of engineering, he remains guided by our animalistic responses to the landscape, which he calls "corporal

→ Bureau Bas Smets,
Landscape Project,
Luma Arles, France, 2021,
Photo: Iwan Baan

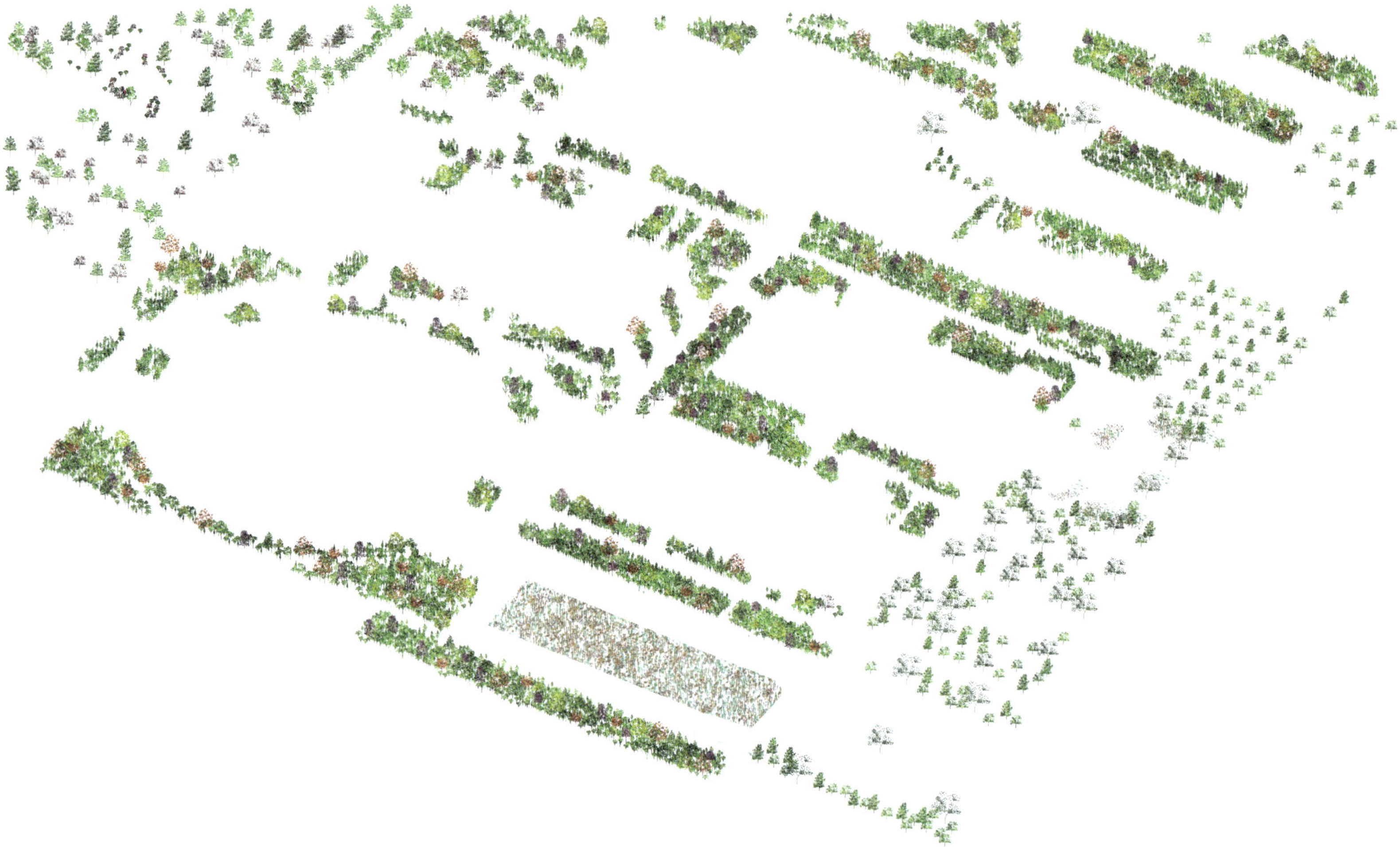

↑ Bureau Bas
Smets, Vitra Campus
Landscape Study, Weil
am Rhein, 2023

← Aerial view of the
Vitra Campus in Weil
am Rhein, 2015

I don't like the word "greening". We're not changing the colour.

Bas Smets

↗ Bureau Bas Smets, Landscape redevelopment of the area surrounding Notre-Dame, Paris, 2022 (rendering)

knowledge". This openness to the symbiosis between analysis and intuition means that he is unafraid to tweak established methodologies, such as so-called "sponge cities". Smets prefers to call this "biospheric urbanism" – an approach he says operates "at the interface between meteorology and geology". By stripping away the dissonant, impregnable layers that have been added to cities over the last century or so, Smets and his team unearth the fertile topography that lies beneath, maximizing its inherent potentialities. That means reactivating, for example, the mutually-propagating cycle that operates between rainwater and plants. At the same time, Smets employs his own iteration of the Miyawaki method – a kind of expedited Darwinist formula that forces plants into competition with one another, accelerating their development and strength. These and other methodologies feature in Smets's study for landscaping the Vitra Campus in Weil am Rhein, where Swiss furniture manufacturer Vitra has, over time, developed an ensemble of production buildings, museums, and architectural icons. In his concept, Smets radically reimagines the site based on a two-pronged strategy. One component

utilizes the water-storage and usage possibilities suggested by the on-site warehouses and other buildings, while the other introduces vegetation to reframe the architectural monoliths that sit within Vitra's aegis. By removing as much as possible of what he refers to as hardscape – sealed surfaces, roads, and other redundant infrastructure – and replacing it with three key softscape typologies – a park, a garden, and a forest – Smets's vision proposes an immersive experience for visitors that responds to and transforms the lived landscape.

"I don't like the word 'greening'. We're not just changing the colour. It does that too, but that's not the main action. For each man-made climate a comparable condition in nature can be searched for and studied. Using its logic, vegetation can be introduced to transform our cities into complex urban ecologies, capable of producing new microclimates." ●

Based on her many years of global research, JULIA WATSON believes that some of the most promising technologies to help tackle future climatic challenges have existed for centuries. Her mission is to introduce this ancient wisdom into contemporary landscape architecture practice.
by Lisa Dabscheck

"We have to shift the way Indigenous knowledge is valued."

The New York-based Australian researcher, landscape and urban designer Julia Watson's book *Lo—TEK: Design by Radical Indigenism* proposes an alternative, arguably even controversial approach to the established discourse on landscape architecture. What if, instead of plunging further into high-tech, foward-focused thinking, sustainable futures could leverage the wealth of age-old Indigenous knowledge? Watson's research led her to nature-based infrastructure projects in 18 countries across the globe, from the Jingkieng Dieng Jri Living Root Bridges created by the Khasi tribe in northern India to the Totora Reed Floating Islands in Peru, bringing her into direct contact with empirical strategies that she believes can powerfully – and sensitively – respond to the climate challenges of our time. Far from being site-specific, Watson suggests, these ancient innovations can be incorporated into contemporary urban landscape projects, creating a powerful nexus between high-tech and "Lo—TEK", a term coined by Watson to reframe value judgements

associated with these technologies. "TEK" stands for Traditional Ecological Knowledge: "We really have to shift the way that Indigenous knowledge is valued in contemporary society", says Watson.

Yet there remains something of a dissonance between the collective appetite for the book's thesis and its translation into live projects. Watson cites various structural and institutional factors for this hesitancy, but remains optimistic: "If we can keep working on some of the limitations for the use of the knowledge and using Lo—TEK as a platform to talk about the incredible opportunities in projects [in symbiosis] with climate change, then we'll have a trickle-down effect. The United Nations will not just be giving billions of dollars towards hi-tech, carbon credits and climate conservation, but to the communities who have actually [developed] these technologies." This, Watson argues, could engender "a change in the landscape in how these technologies can scale and be used".

→ Acadja Fish Paddocks in Ganvie, Benin. The town in the south of Benin floats on Lake Nokoué, surrounded by 12 000 Acadja fish ponds, which act as artificial mangroves that offer refuge to many other species. Photo: Iwan Baan

↙ Illustration of the 20-year pyrotechnology ecology of the Maya Milpa Cycle. The Milpa Cycle has been practiced in Mayan forest gardens for more than 8000 years. During the cycle, burnt land slowly transforms back into a forest with the help of annual vegetable crops.

Illustration from Julia Watson, *Lo—TEK: Design by Radical Indigenism*, 2019

0 yr

STAGE 4
Crop forests grow into hardwood forests and the milpa cycle begins again with burning

STAGE 1
Forests are planted with the Three Sisters

5 yr

STAGE 2
The Three Sisters are planted with quickly yielding fruit trees to create forest gardens

STAGE 3
Forest gardens are planted with slow yielding fruit trees to create food forests

15 yr

10 yr

I've been waiting for one of these big design studios to say, "Okay we have the capacity, capability, expertise, time, clientele, and budget to be able to pull these ideas into a project. These are innovative, these are necessary, we need to decolonize industry, we need to work with climate change."

Julia Watson

192

Adding to the complexity is the question of bringing these technologies into a contemporary socio-political landscape that is charged with concerns about the legacies of colonialism and the risks of cultural appropriation. "[This] has come up over and over again, and rightly so", Watson acknowledges. "The response is allyship [...] trying to put in best practices informed by Indigenous communities. [...] So working together, getting permission from communities to work with them. [...] to actually publish any of this information." For a project entitled *Our Time on Earth* at the Barbican Centre in London, Watson connected engineering giant Buro Happold with leaders from three Indigenous communities, in the process developing a legally-binding oral contract known as a Smart Oath of Understanding (SOOU). "It is actually written in to the block chain and so it's transparent in perpetuity", says Watson. "It talks about the value of this knowledge,

the way this knowledge will be used for the benefit of humankind. [...] how to reimburse for this knowledge, and then the communities responded to us in their Indigenous language to accept the contract."

The establishment of due processes and documentation for these technologies, Watson argues, forms a modus of knowledge-keeping that can be essential to the survival of Indigenous communities' intellectual property. At the same time, global awareness can add enormous perceived value that, crucially, can protect certain communities from the very real dangers posed by outside forces, such as governments, to their knowledge and environments. Together with her field-based research, Watson runs a Brooklyn-based design studio strongly founded on replacing prevalent individualized systems of planting with collective nature-based practices such as rewilding that are sustainable, scalable, and respond to a shared global ethos to create an interconnected, reciprocally-resilient planet ●

↓ Illustration of the process of wastewater treatment at East Kolkata's bheris wetlands. The facility is operated by a fishermen's cooperative.

Illustration from Julia Watson, *Lo—TEK: Design by Radical Indigenism*, 2019

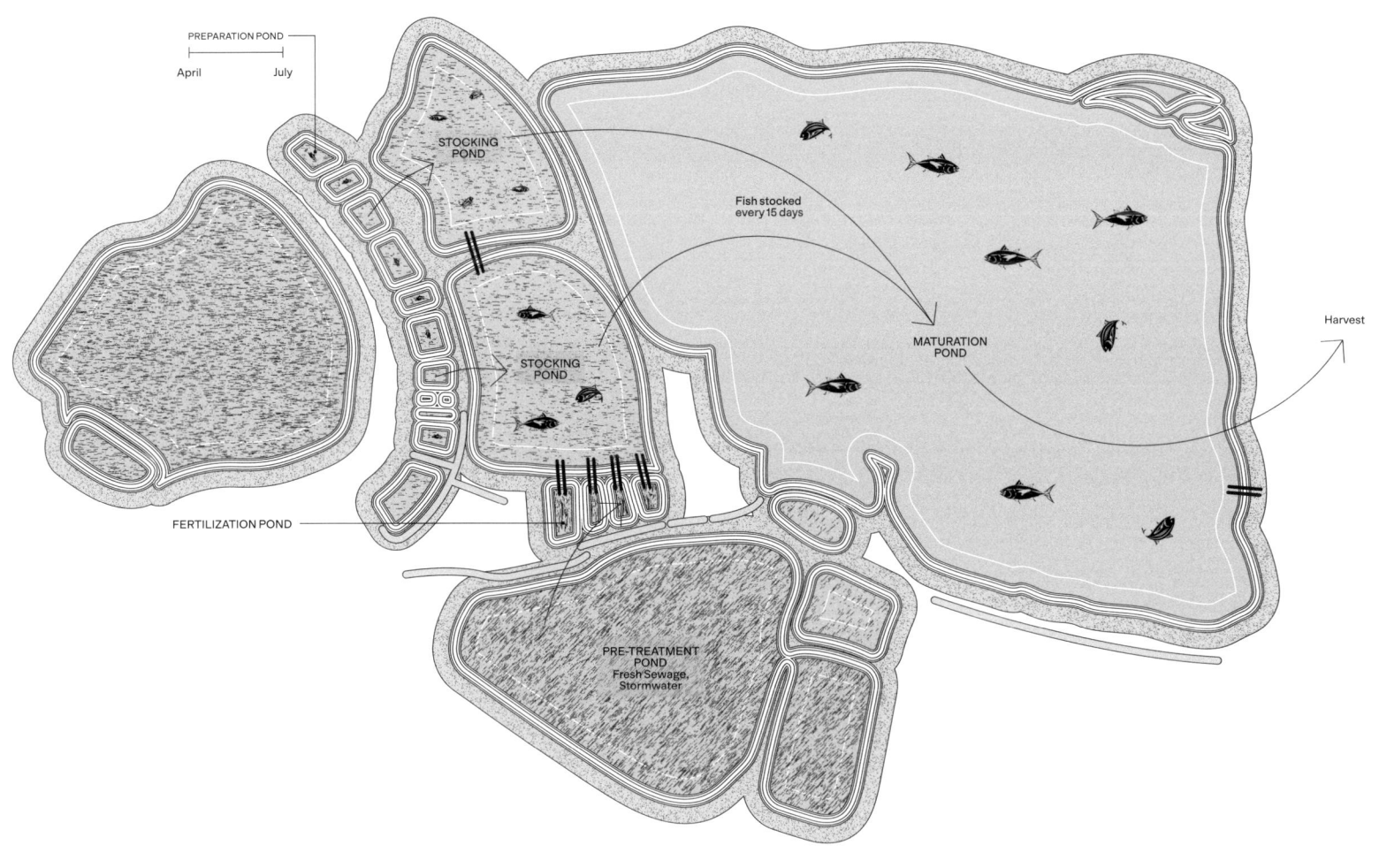

A Garden
of Ideas

When it comes to future challenges for ecological
and social sustainability, the answer may be found
in the garden and in the act of gardening. Maria
Heinrich, Marten Kuijpers, Viviane Stappmanns,
and Lisa Dabscheck glean a collage of ideas
around how to collaborate with each other and with
nature in our cities, buildings, schools, and living
environments. The collective picture is one of hope.
And just like in the garden itself, there is no one-
size-fits-all solution. Trial and error, responding to
local conditions, and the realization that everything
is connected are always at play.

Growing Objects

In the mid-2000s, when 3D prototyping and printing became viable technologies in design and manufacturing, Alice and Gavin Munro took an alternative route. They reconsidered human interaction with nature by looking back to the agricultural revolution, and as a result decided to grow furniture – literally dispensing with conventional production altogether to instead use "air, soil, and sunshine" as source materials in a process they call designing with a "carbon sink 3D printer". In 2008, they planted 3000 trees. It took four years and much trial and error before the first harvest of prototypes (chairs, lamps, and other experiments) was produced, via "biofacture" as opposed to manufacture, in 2012. In the years since, the couple have refined their products and established Full Grown, an organization dedicated to researching, testing, prototyping, and putting into practice agricultural design methods to shape trees directly into objects. These – mostly chairs and lights – can take up to nine years to grow, and each is shaped individually as it matures. No two pieces are alike, with different species revealing their peculiarities as they develop. While the initial plan to harvest a chair was to grow four trees next to one another and join them to form a seat, the current crop are trees grown upside down around a custom-made frame.

→ Full Grown, The Grown Chair, 2012–16

←↓ The Full Grown orchard in Wirksworth, Derbyshire, where up to one hundred objects are maturing at any given time.

Designing Empathy

The effects of climate change are usually felt in seemingly faraway locations, creating a sense of emotional as well as physical distance. Through her large-scale, soft-textured, interactive works, Argentinian-born textile artist Alexandra Kehayoglou evokes feelings of empathy and care for vulnerable landscapes that face dramatic threats in the years and decades ahead. In doing so, she seeks to find an immersive way to call for environmental preservation and awareness, inspiring those who interact with her works to think of the planet as a common garden that needs to be collectively tended. Hailing from a carpet-making family of Greek descent, she has applied traditional hand-tufting techniques to recreate landscapes such as the Santa Cruz River in Argentina, where her work conjures the beauty of the last free-flowing wild river in the country, a natural wonder that is now slated to be the site of two major hydroelectricity dams. Her latest exposition centres on the Greek island of Milos. The landscape of the small island continues to be dramatically altered by activity in over 200 mines, most of which are used to unearth rapidly vanishing rare minerals. Yet once a year for a short period the barren landscape continues to be the site for an abundant blossoming of wild-flowers, an occurrence that in this context could be read as the earth reasserting itself.

↙ Alexandra Kehayoglou, *Santa Cruz River*, 2016–17, tapestry, Collection National Gallery of Victoria, Melbourne

→ Alexandra Kehayoglou, *Santa Cruz River*, 2016–17, tapestry, at National Gallery of Victoria, Melbourne

↓ Alexandra Kehayoglou, Impression from research on the Greek island of Milos, 2022

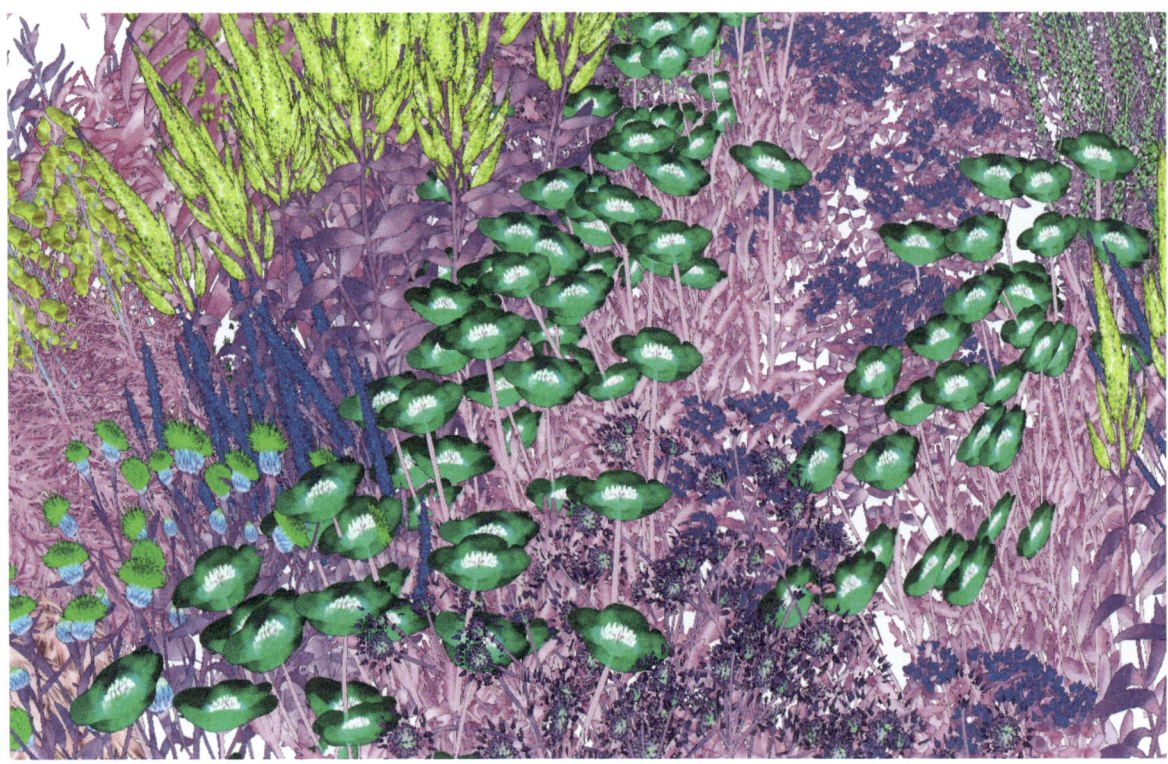

Art for Pollinators

If bees, butterflies, moths, wasps, and other pollinating insects could design their own gardens, what would humans see? Certainly something very different from the gardens we design for ourselves. For the past decade, British artist Alexandra Daisy Ginsberg has created artworks across various scales to critique and question the human-centricity of design – from the perspective of tiny, synthetically engineered organisms to a digitally revived extinct rhino. Her project *Pollinator Pathmaker,* "an artwork for pollinators, created and cared for by humans", inverts the usual imperatives and economies of scale and scarcity. With the goal of becoming "the world's largest climate positive artwork", *Pollinator Pathmaker* aims to counteract the decline in flying insect numbers, which in Germany alone have plummeted by 75 per cent in 30 years. Ginsberg convened the brainpower of horticulturists, scientists, and pollinator experts into an altruistic algorithmic tool and encouraged thousands of grassroots participants to collaborate on creating this decentralized living artwork for pollinators – in suburban gardens, in schoolyards, on roofs and balconies. Originally commissioned by the Eden Project in Cornwall, UK, *Pollinator Pathmaker* begins with an interactive website, pollinator. art, where participants feed in the parameters of their garden-to-be. Once location, climate, size, and soil specifications have been entered, the *Pollinator Pathmaker* generates a unique garden plan optimized to serve the maximum diversity of pollinators. Users can print their planting instructions to help them create their piece of the *Pollinator Pathmaker* puzzle and receive a digital certificate for their artwork. From the first large "Edition" gardens, which opened in 2022 at the Eden Project and for London's Serpentine Gallery, the project is now being catapulted from the UK into new climatic zones. As this requires expanding the plant palette, Ginsberg collaborates with local experts and art institutions. In Germany, Light Art Space and the Naturkunde Museum Berlin, in 2023, are creating a large public garden made for pollinators and providing a catalyst for many more, smaller "DIY Editions" across the city, each supporting the flourishing of the others.

I want to transform how we see gardens
and who we make them for: let's create art
for pollinators' tastes, not human taste.

Alexandra Daisy Ginsberg

↖ Alexandra Daisy Ginsberg,
Pollinator Pathmaker, 2022,
digital rendering

← *Pollinator Pathmaker*,
Eden Edition Garden in Bodelva,
Cornwall, photographed in
July 2022

↗ Alexandra Daisy Ginsberg,
Pollinator Pathmaker, online tool

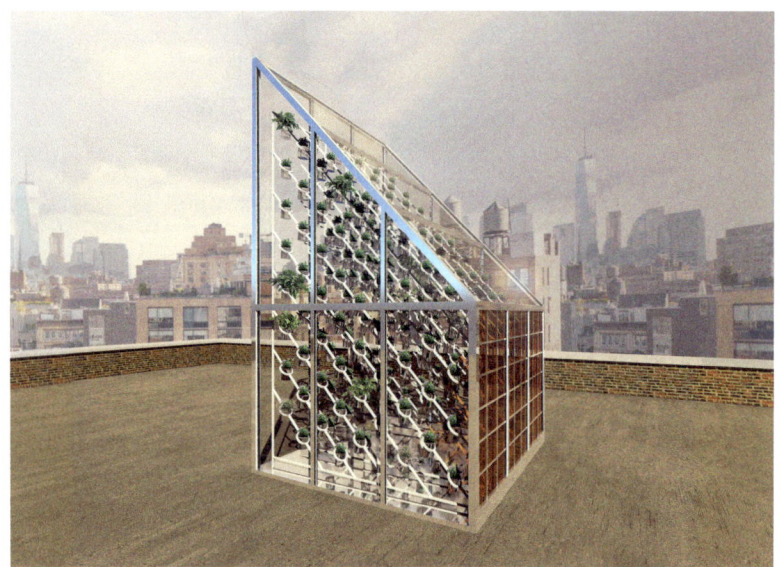

Power Plant

More than half of the world's population today lives in urban areas. According to United Nations' predictions, this will increase yet again by nearly 20 per cent. Crises and disruptions to supply chains make urban populations increasingly vulnerable to food insecurity. And as cities grow so do the transport routes for fresh food. Dutch designer Marjan van Aubel tackles the twin problems of energy shortage and food insecurity with her Power Plant design, which allows for food cultivation on rooftops or other built-up structures so food can be produced at the site of consumption. The greenhouse structure uses solar energy to power a hydroponic feeding system for plants, saving water usage by up to 90 per cent compared to traditional soil farming. It was on a train journey through the Westland in her native Netherlands that van Aubel was struck by how 80 per cent of the cultivated land in this region consists of greenhouses. "There I saw all this glass and thought, what if we integrate those with transparent solar panels?" Her next thought was how effective it would be to place smaller, solar-powered greenhouses on urban rooftops, thereby utilizing empty spaces in the heart of the city to grow food.

↖ ← ↓ Marjan van Aubel, Power Plant (rendering), 2018, designed in collaboration with the Nieuwe Instiuut, Rotterdam, architect Emma Elston, researcher Yasmine Ostendorf, Physee, and the University of Amsterdam

There is widespread concern that the risk of food shocks – sudden disruptions to food supply – is increasing. It emerges that a city's vulnerability to food shocks can be reduced by diversifying its supply chains.[1]

Zia Mehrabi, Assistant professor, Sustainability Innovation Lab, University of Colorado

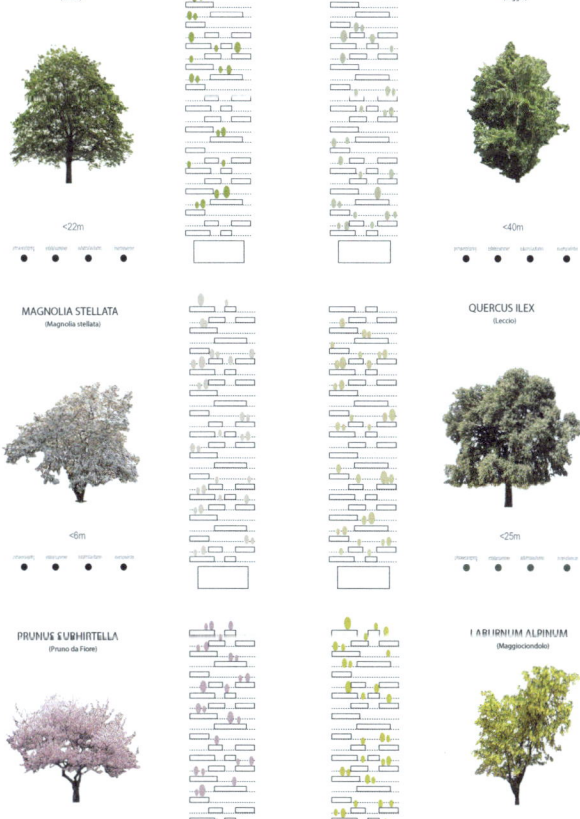

Vertical Forestry

The 400 terraces of Stefano Boeri Architetti's twin high rises in Milan's "Bosco Verticale" – or vertical forest – house around 800 trees, 4500 shrubs, and 1500 other plants on approximately 1700 square metres of space, all of it nourished with filtered waste water. Planting took over a year, but even more complex was the research that went into identifying the right species to withstand the conditions on the 111-metre-tall tower. Botanist Laura Gatti chose 20 tree and 80 plant species, among them Ash, Ilex, and olive saplings. The trees were specifically grown and "trained" for their life on the tower, with its compact spaces and maximum environmental exposure. In turn, the vegetation acts as a natural air-cleaning system to counteract pollution, and functions as a visual metaphor that links urban and natural ecosystems. Since completing the building in Milan, further iterations have been created or planned elsewhere. Among them is a social housing tower in the Dutch city of Eindhoven and a tower featuring Mediterranean natives such as rosemary in the Albanian capital of Tirana. All utilize local plant species.

↑ Boeri Studio, Bosco Verticale, Milan, 2014

→ Diagram of trees for Bosco Verticale, 2014

The plants have been selected by the botanical consultants Emanuela Borio and Laura Gatti.

More Outside Inside

Domestic gardens can be seen as spaces for the privileged.
The notion of the winter garden takes this even further,
conjuring up images of dedicated suntraps in expansive
dwellings. But what if winter gardens – essentially glassed-in
rooms swathed in light – could be affixed to the homes of
the less privileged instead of demolishing the building and
replacing it with a larger one? This was precisely the concept
that French architects Anne Lacaton and Jean-Philippe
Vassal brought to three blocks of a 1960s' housing estate on
the outskirts of Bordeaux, France. Like much social housing
of the era, this estate was built using a typical template and
was surrounded by 20 hectares of mostly unused green
space. With their project, Lacaton and Vassal brought private
sanctuaries directly to the people, creating hybrid spaces
that connect the inside to the outside. Affixed to the building
facades like an outer skin, the winter gardens provide
spaces that afford expansive views over the surrounding area,
along with balconies that also comprise part of the project,
which altogether cost roughly half as much as a new build.
The project increased the floor space for the residents
twofold with minimal disruption to their lives, and gave them
autonomy as to how to use their new conservatories.

↗ Lacaton & Vassal, Druot, Hutin, Transformation of apartment
buildings at the Cité du Grand Parc, Bordeaux, 2017

↘ 530 dwellings were extended with winter gardens.

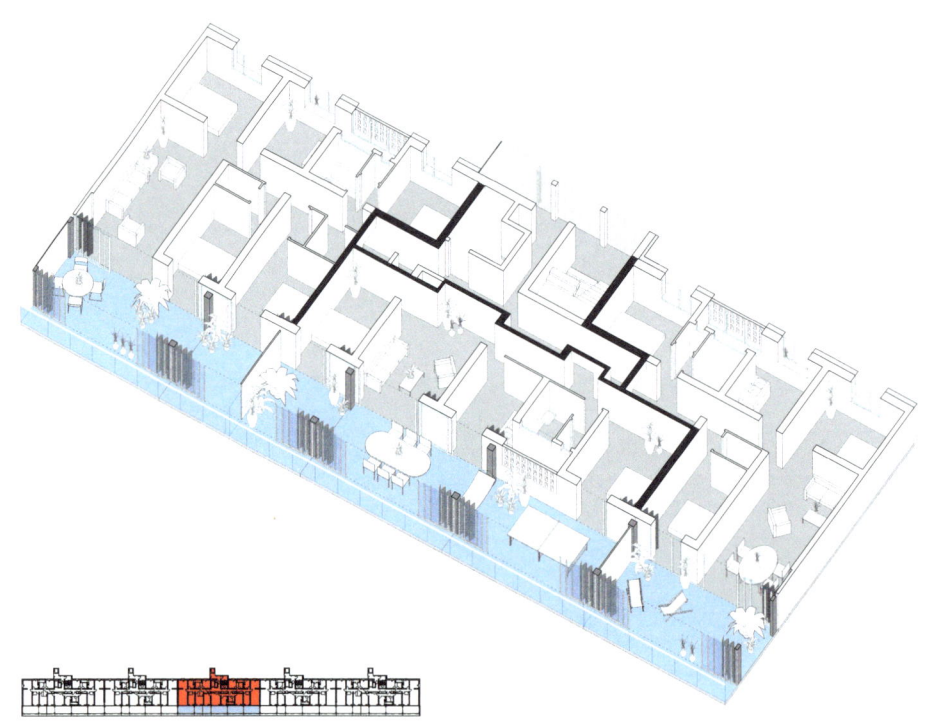

The World Health Organization recommends […] a minimum of 9 square metres of green space per individual, with an ideal urban green space (UGS) value of 50 square metres per capita.[2]

Alessio Russo and Giuseppe T. Cirella, Researchers

Potted Gardens

A city of almost 14 million inhabitants, Tokyo only features around five square metres of green space per person. In comparison, New York City, bolstered by its 3.5-square-kilometre Central Park, has 26.4 square metres for each inhabitant. As a result of this scarcity, there is a long tradition of Tokyo residents placing small, half-private, half-public flowerpot gardens in whatever space they can find. Architect Ryue Nishizawa picked up on this when he designed Garden & House, a five-storey dwelling that covers a floor with an area on each level of only eight by four metres. Giving prominence to an abundance of plants, which are arranged in various soft formations that reach out to the periphery on every level, the building dispenses with conventional hard borders of inside/outside, giving the impression that it is devoid of walls. In fact, recessed glass partitions provide privacy and security. But these are set back from the facade and conceal the office spaces, bedrooms, living room, and bathroom, so that the building appears to be populated only by foliage.

↗ Ryue Nishizawa, Garden & House, Tokyo, 2006–12

Gardening by Numbers

What are urban gardens actually worth? In the capitalist system that judges the value of all things according to economic parameters, the benefits of gardens and their attendant ephemeral qualities can be easy to grasp but difficult to measure. Funded by the Federal Ministry of Education and Research, scientists at the German Institute for Ecological Economy Research have nevertheless tried. The result: of Berlin's nearly 3000 hectares of garden area, vegetables and herbs are grown on about 140 hectares, or just under five per cent, yielding about 7600 tons of produce. This covers the annual needs of 50,000 people and has a financial value of about ten million euros. If, hypothetically, a quarter of the city's population of four million wanted to follow suit, almost all of Berlin's parks and public and private gardens would need to be used for growing food. Because space is an issue, the scientists went on to develop a space-saving vertical garden. Fed by 1000 litres per day of grey water from showering and using 36 times less space than a conventional garden, a prototype vertical farm produces 160 kilograms of salad per year, enough for 28 people. If this was extrapolated to all of Berlin, then only 26 hectares of space (just under 20 football fields) would be needed to provide all Berlin residents with 5.8 kilograms of salad per year (the current per capita consumption), in comparison to 836 hectares required by conventional farming (nearly 600 football pitches).

→ Vertical farming module "Himmelbeet-Tower", supplied with rainwater from the roof of the garden café

↘ Aerial photograph of the community garden Himmelbeet, Berlin-Wedding, 2021

Growing Together

What would a socially responsible garden look like? One version might be a green space that fosters social cohesion, embraces biodiversity, and reduces food waste while providing food security, as well as incorporating youth education and contributing to health and well-being. That's the ambitious yet highly successful – and easily replicable – concept behind Edible Estates, a Scottish not-for-profit organization that converts public land, predominantly around social housing estates, into community gardens, where residents can literally eat the rewards of their horticultural efforts and help to invigorate a sense of shared goals with their neighbours. Having seen that large areas of grassland around social housing are considered a maintenance problem, Edible Estates have worked with residents to develop them as community assets. They have developed an open source design, providing a blueprint for those seeking to replicate their approach. These neighbourhood gardens – each of which is equipped with a communal shed – provide a focus for community coalescence, building cohesion. This is an exemplar of how applied, design-led practice can engender meaningful change. As a result, local councils have been made more aware of implementing place-making strategies, such as the 20-Minute City, where residents can access all their needs without the use of a car.

↗ Edible Estates, Greenway Garden in Edinburgh's low-income neighbourhood of Wester Hailes, 2022

Since beginning with a single school site in 2001, the Stephanie Alexander Kitchen Garden Program now supports over 1000 early-years childhood services and primary and secondary schools across Australia.

Learning to Grow

After a prolific career as a restaurant owner, chef, and cookbook author, Stephanie Alexander turned her attention to primary schools in order to educate children from a young age about the rewards of being actively involved in their own organic food production and consumption. In her native Melbourne, Alexander began a partnership with a local school to set up a garden and kitchen for the students, where they learned about planting, tending, harvesting, cooking, and composting as part of the school's curriculum. The broader objective was to encourage healthy eating and avoid childhood obesity. Three years after her first foray, she established the Stephanie Alexander Kitchen Garden Foundation to support schools around Australia by giving advice and offering consultation on setting up their own kitchen garden programmes. The Stephanie Alexander Kitchen Garden Foundation helps with the set-up, from how the kitchen needs to be organized to planting advice, as well as sharing recipes and supporting curriculum integration, all with the mandate to build social cohesion and community resilience.

↑ Garlic harvest at Moonah Primary School, Derwent Park, Tasmania, 2020

↗ Bush Tucker Garden at the Junction Park State School, Annerley, Brisbane, Queensland, 2020

Edible Neighbourhoods

When local residents of the new neighbourhood of Rijnvliet in Utrecht, Netherlands, were given the opportunity to suggest how they would like their region to be environmentally enriched, they came up with an unusual proposal: they wanted a local forest in which they could grow, harvest, and ultimately eat their own produce. The municipality embraced their initiative and engaged a triumvirate of experts to design the public spaces in the area as edible, spirit-enriching, and educative landscapes. The residents effectively became "co-designers" when their ideas were integrated into the design phases of what has become the Rijnvliet Edible Neighbourhood, an award-winning approach to sustainability, education, and edibility. Rijnvliet extends over a 15-hectare integrated ecosystem comprising more than a thousand fruit trees and 220 subspecies of edible plants, herbs, and shrubs which symbiotically reinforce each other – and all of which are edible for humans and animals. As well as providing community-building and self-sustaining elements, Rijnvliet, with its central food forest, is designed to augment water management, heat reduction, and air purification.

↖ De Zwarte Hond (master plan), Felixx (landscape architecture) and Æ Food Forestry (food forest planning), Rijnvliet Edible Neighbourhood, Leidsche Rijn near Utrecht, 2021

↙ Diagram of animals inhabiting the Rijnvliet Edible Neighbourhood

I am interested in the idea of "architect as gardener".
Someone who's there strategically, every single day.
Paying attention and listening and watching and
cultivating and caring for things as opposed to imposing
a vision and then forcing people to live in in it.[3]

Fritz Haeg

Cultivating Community

Can you grow a community like a garden? This is the question that the revolving group of Salmon Creek Farm residents explore through art projects, land care, programmes, and much more. Just two miles inland from the Pacific Ocean in northern California and encircled by a second-growth redwood forest, Salmon Creek Farm was initially established in 1971 by a group of young people who rejected mainstream culture and were part of a global back-to-the-land movement that emerged from the student protests of the late 1960s. Artist Fritz Haeg purchased the farm in 2014 to continue this legacy, forming a new chapter in its history as a long-term art project shaped by many hands, a sort of queer commune-farm-homestead-sanctuary-school hybrid. In 2022, a non-profit organization was established to run land-based artist programmes for half the year, supported by two-week retreat rentals for the other half. Located on Central Pomo land, the community aims to cultivate, reconnect, and weave back strands of Indigenous environmental knowledge.

↑↗→ Salmon Creek Farm, Mendocino Coast, California, since 1971 and again since 2014

The Forest as a Garden

Today, food forestry is becoming more and more popular, but its pioneer was Robert Hart. He was an English horticulturist who developed a method of forest gardening for temperate zones. After experiencing the extreme labour that annual plants require, he experimented with perennial plants at his farm Highwood Hill in Wenlock Edge, Shropshire. As he was working the farm on his own and taking care of a less-abled brother on the side, the forest garden soon became the brothers' livelihood. They lived off a 500-square-metre organic forest, adopting a vegan and mostly raw food diet. Hart's method consists of seven layers: canopy, low tree, shrub, herbaceous, ground cover, rhizosphere, and vertical. Until his death in 2000, he promoted this gardening approach in publications and interviews. He wanted to encourage others to start forest gardens in their backyards as a tool to regreen cities and to make people more self-sufficient and healthier.

↘↘→ Robert Hart's Forest Garden at Wenlock Edge, Shropshire, 2017

↓ The seven layers of a forest garden according to Robert Hart (illustration by Graham Burnett)

1. Canopy (large fruit and nut trees)
2. Low tree level (dwarf fruit trees)
3. Shrub layer (currants and berries)
4. Herbaceous layer
5. Soil surface (ground cover crops)
6. Rhizosphere (roots and fungi)
7. Vertical layer (climbers and vines)

The Seven Layers of the Forest Garden

A 1000-Year Idea

Corporate success can lead to complacency about the natural world, even when such success comes directly at its expense. But, in some cases, there is an impetus to create a meaningful remedy. This was the compulsion for Mitsushige Hayashi to offset the carbon footprint of his lucrative national newspaper business. Tokachi Millennium Forest is a conservation project set in the wooded foothills of the Hidaka Mountains on Japan's northernmost island of Hokkaido. With a 1000-year sustainable vision, the venture aims to provide an antidote to the diminishing natural habitats on the island, as well as to offer a green retreat for Japan's predominantly urban population and opportunity to interact with the landscape that is the source of their ancient cultural identity. The forest consists of three central elements: an Earth Garden with integrated restaurant and undulating grassland topography that evokes the ebb and flow of the ocean and draws visitors out from the eatery's familiar comforts to engage with the landscape; an ornamental Meadow Garden, a congregation of colourful perennials; and the Entrance Forest, where native species have been replanted in their natural territory. Tokachi Millennium Forest is criss-crossed by boardwalks and paths that bring visitors into direct contact with the textures, scents, colours, and sounds that encourage them to form deep connection with the natural world. Key to the beauty evoked by the forest is the authenticity of its planting, conceived by the British designer Dan Pearson according to the Japanese culture of *satoyama*, which he describes as "about living closely with the land and only taking as much as you need from it", and sensitively nurtured by head gardener Midori Shintani. "*Sato* means village. *Yama* means mountain", she explains. "I am especially keen to learn from Japanese tradition and harmonize our mother culture and our modern gardening."

→↘↓ Dan Pearson (design) and Midori Shintani (care and maintenance),
Meadow Garden at Tokachi Millennium Forest, 2004–08

Japanese nature worship is to live with sincere devotion to nature. [...] This spirit, inherited from our ancestors, is surely alive in our forest.[4]

Midori Shintani, Head gardener,
Tokachi Millennium Forest

Post-Extractivist Landscapes

Louvre-Lens Museum is situated in a 20-hectare park designed by landscape architect Catherine Mosbach in northern France. This subsidiary museum of the Louvre in Paris is placed on the grounds of a former coal mine in the industrial city of Lens. Mosbach's design interprets this post-extractivist landscape as a symbiotic environment, where soil, water, organisms, as well as pollution form an ecosystem. As such, it is a healing site where the memory of the economic and ecological devastation of the ground is decisive for the composition of soil and plants. The post-industrial park is tended to by four gardeners, whose task is to preserve the biodiversity of the site and create a space of comfort in an industrialized environment.

↑ Aerial view of the urban center of the city of Lens, the university town and the mining basin of Nord-pas-de-Calais, January 1969

↖ Mosbach Paysagistes (landscape architecture), SANAA (architecture), Passage sous Bois Pionnier (Path between the pioneer trees) and Prairie Estrade-Auditorium, Louvre-Lens Museum Park, 2012

Post-Plantations

CATPC (Cercle d'Art des Travailleurs de Plantation Congolaise) is an art cooperative of plantation workers in Lusanga in the Democratic Republic of the Congo. Their aim is to buy back the depleted land of former plantations by making and selling art, turning them into ecological, sustainable, multi-species post-plantations. The plantation workers create sculptures out of clay, which are then digitally scanned and remade from plantation materials such as cocoa, sugar, or palm fat, and shown in museums worldwide. Alongside the sculptures, they also create NFTs as a tool for digital restitution, criticizing the current ownership of many cultural artefacts of the African continent and the conse-quences of colonialism. As of 2022, the farmers have bought 50 hectares of land and transformed it into a food forest with a museum on-site. This museum, White Cube, was designed by OMA and is a centre for a museum programme aimed at decolonizing the plantation and creating the post-plantation.

↑ Agronomist Charles Munanga with cocoa seedlings in CATPC's plant nursery, Lusanga, 2017

↗ CATPC's plant nurseries in the post-plantation, Lusanga, 2017

→ Mbo Mangala with cocoa seedlings in CATPC's plant nursery, Lusanga, 2017

The Gardens of the Great Green Wall

The Great Green Wall was initiated by the African Union in 2007 to regreen the Sahel region and limit desertification. This 8000-kilometre-long strip south of the Sahara is threatened by the spreading desert, with repercussions for inhabitants and the climate of the whole planet. Every affected nearby country uses different methods that are often based on Indigenous practices to support the Great Green Wall. One of these are the Tolou Keur gardens in Senegal. They are circular food forests consisting of different sections of heat-resistant plants, beginning with medicinal herbs in the centre of the circle, to fruit, vegetables, and even trees for timber production on the outer rings. Through the circular shape of the garden the roots of the plants grow towards the centre and moisture is retained in the rhizosphere. These gardens can not only stop the encroachment of the desert, but also make communities self-sufficient.

↓ Tolou Keur gardens, Boki Diawe, Matam region, Senegal, 2021

Coastal Care

Sato-umi describes a state of a landscape found on the southern coasts of Japan where biodiversity in fact increases because of human interaction. This is owed largely to a community of female breath-hold divers, the Ama, who forage abalone, seaweed, and pearls. Their name can be translated as "women of the sea". They have been diving for more than 2000 years, forming a vital connection between the land and the sea; the Ama are a matriarchal community who develop a relationship of care and respect for the sea by exercising a practice of minimum intervention. Their regenerative ethics are closely related to that of the Haenyeo, female divers on Jeju Island in South Korea, who dive for seafood. Both diving cultures are known for resilience in adversity and women's independence; however, the number of women divers has decreased drastically in recent years.

Photography by Nina Poppe from the series *Ama*, 2010

↑ In the huts on the beach, the Ama store their equipment, eat together, and warm themselves by the fire after diving.

↗ Some divers start their dive from land, others go out to sea in groups by boat.

→ The almost 80-year-old diver returns to the boat after a 90-minute dive. The Ama dive without equipment. The white headgear is supposed to protect them from sharks.

Floating Heritage

Chinampas are floating agricultural gardens on the lakes and swamps of the Valley of Mexico, a place on the outskirts of the country's capital. This construction method dates back to Mesoamerican times when the Aztecs built the foundation of their capital Tenochtitlan on artificial islands. As all agricultural cultivation and gardens were built as Chinampas, these hydraulic and engineered constructions formed the basis of the Empire's wealth. The Spanish colonizers later destroyed and drained many of the lakes, and so only some Chinampas are still cultivated in present-day Mexico City. This hydroculture method is especially prosperous because of its irrigation and drainage system, enabling cultivation in otherwise unusable land and making the swamps fertile. It has been designated as one of the Globally Important Agricultural Heritage Systems by the United Nations.

↑ Modern-day Chinampa in Xochimilco, south of Mexico City, 2019

The End of the Green Desert

"Plants are the basis of all terrestrial biodiversity", states James Hitchmough, Professor Emeritus of Horticultural Ecology at the University of Sheffield, where, over a long career as an academic, he developed a prolific practice as a plantsman and landscape gardener. He has a particular interest in reducing the proliferation of "green deserts" – this is how he refers to mowed lawns – in urban environments. "It seems a little bit perverse to cover 25 per cent of cities with something virtually nothing lives in", he suggests. While he acknowledges that lawns have their uses, he feels that there is a mandate to scale back and replace the lawn with something more multifaceted and alive. If we reduce our "green deserts", how would Hitchmough like our urban environments to change in the next 50 years? "I think we need more complexity and diversity", he replies. And what would we replace mown grass with? "I am interested in something we call wood pasture. It's like an orchard with a meadow beneath it. So you can still see the sky." Professor Hitchmough's vision takes pastoral elements and brings them into urban landscapes, an innovation he admits is "a bit of a clash with the prevailing aesthetic of modernism", but one that would populate our lived environment with species that reflect the multiplicity of our planet.

↘ James Hitchmough, Wildflower Meadow, Chatsworth House, Derbyshire, May 2022

1 Zia Mehrabi, "How to Buffer against an Urban Food Shortage", *Nature* (7 July 2021), p. 175.

2 Alessio Russo and Giuseppe T. Cirella, "Modern Compact Cities: How Much Greenery Do We Need?", *International Journal of Environmental Research and Public Health*, vol. 15, no. 10 (2018), online: https://www.mdpi.com/16604601/15/10/2180, accessed 20 January 2023.

3 Fritz Haeg, Initiator of Salmon Creek Farm Community (since 2014), in an interview with Alice Rawsthorn, on *Design Emergency*, 18 March 2021.

4 Midori Shintani quoted from Dan Pearson with Midori Shintani, *Tokachi Millennium Forest – Pionieering a New Way of Gardening with Nature*. London: Filbert Press, 2021, p. 48.

Annex

Aben, Rob, and Saskia de Wit, *The Enclosed Garden: History and Development of the Hortus Conclusus and its Reintroduction into the Present-day Urban Landscape.* Rotterdam: 010 Publishers, 1999.

Buckminster Fuller, Richard, *Operating Manual for Spaceship Earth.* Carbondale: Southern Illinois University Press, 1969.

Carson, Rachel, *Silent Spring.* London: Penguin, 2000/1962.

Clément, Gilles, *"The Planetary Garden" and Other Writings.* Philadelphia, PA: University of Pennsylvania Press, 2015.

Clément, Gilles, *Jardins, paysage et génie naturel.* Paris: Collège de France/Fayard, 2012.

Clément, Gilles, *Manifesto of the Third Landscape.* Trans Europe Halles, 2022, published online: https://bit.ly/ThirdLandscapeManifesto

Cluitmans, Laurie, *On the Necessity of Gardening: An ABC of Art, Botany and Cultivation.* Amsterdam: Valiz, 2021.

Coccia, Emanuele, *The Life of Plants: A Metaphysics of Mixture.* Cambridge, UK/Medford, MA: Polity Press, 2019 [Fr. original: *La vie des plantes: Une métaphysique du mélange.* Paris: Editions Payot & Rivages, 2016].

Corner, James (ed.), *Recovering Landscape: Essays in Contemporary Landscape Theory.* Princeton, NJ: Princeton Architectural Press, 1999.

Derek Jarman: A Portrait, with an introduction by Roger Wollen. London: Thames & Hudson, 1996.

Derek Jarman's Garden, with photographs by Howard Sooley. London: Thames & Hudson, 1995.

Doherty, Gareth, *Roberto Burle Marx Lectures: Landscape as Art and Urbanism.* Zurich: Lars Müller Publishers, 2020.

Dulk, Leo den, *Mien Ruys: Tuinarchitect 1904–1999: De complete biografie: zoeken naar de heldere lijn.* Rotterdam: Uitgeverij de HEF Publishers, 2017.

Emilio Ambasz: Emerging Nature, Precursor of Architecture and Design. Zurich: Lars Müller Publishers, 2017.

Giesecke, Annette, and Naomi Jacobs (eds), *Earth Perfect? Nature, Utopia and the Garden.* London: Black Dog Publishing, 2012.

Giesecke, Annette, and Naomi Jacobs (eds), *The Good Gardener? Nature, Humanity and the Garden.* London: Artifice Press, 2015.

Girot, Christophe, *The Course of Landscape Architecture: A History of our Designs on the Natural World, from Prehistory to the Present.* London: Thames & Hudson, 2016.

Girot, Christophe, and Dora Imhof, *Thinking the Contemporary Landscape.* Princeton, NJ: Princeton Architectural Press, 2016.

Helphand, Kenneth I., *Defiant Gardens: Making Gardens in Wartime.* San Antonio, TX: Trinity University Press, 2008.

Hobhouse, Penelope, *The Gardens of Europe.* London: George Philip Limited, 1990.

Hoffmann, Jens, and Claudia J. Nahson (eds), *Roberto Burle Marx: Brazilian Modernist.* New Haven, CT, et al.: Yale University Press, 2016.

Holmgren, David, *RetroSuburbia: The Downshifter's Guide to a Resilient Future.* Hepburn Springs: Melliodora Publishing, 2018.

Hu, Fang, *From Semiotics to Energetics Vol. 1, Zheng Guogu: The Everlasting Garden.* Guangdong: The Pavillon, 2021.

Jarman, Derek, *Modern Nature.* Woodstock, NY: Overlook Press, 1994.

Jewell, Jennifer, *The Earth in Her Hands: 75 Extraordinary Women Working in the World of Plants.* Portland, OR: Timber Press, 2020.

Marot, Sébastien, *Taking the Country's Side: Agriculture and Architecture.* Barcelona: Polígrafa Ediciones, 2019.

McHarg, Ian, *Design with Nature.* New York: Natural History Press, 1969.

Mollison, Bill, and David Holmgren, *Permaculture One: A Perennial Agricultural System for Human Settlements.* Hepburn Springs: Melliodora Publishing, 1978.

Morton, Timothy, *All Art is Ecological.* London: Penguin Random House, 2018.

Mostafavi, Mohsen, and Gareth Doherty (eds), *Ecological Urbanism.* Zurich: Lars Müller Publishers, co-published by Harvard University Graduate School of Design, 2016.

Kassler, Elisabeth B., *Modern Gardens and the Landscape*, exhib. cat. The Museum of Modern Art, New York, 1964.

Keogh, Luke, *The Wardian Case: How a Simple Box Moved Plants and Changed the World.* Chicago: University of Chicago Press and London: Kew Publishing, 2020.

Kincaid, Jamaica, *My Garden (Book).* New York: Farrar Straus & Giroux, 1999.

Olonetzky, Nadine, *Inspirations: A Time Travel through Garden History.* Basel: Birkhäuser, 2017.

Orlow, Uriel, and Shela Sheikh, *Theatrum Botanicum.* Berlin: Sternberg Press, 2018.

Oudolf, Piet, and Noël Kingsbury, *Oudolf Hummelo.* Amsterdam: Rights & More, 2015.

Pearson, Dan, with Midori Shintani, *Tokachi Millennium Forest – Pioneering a New Way of Gardening with Nature.* London: Filbert Press, 2021.

Pih, Darren (ed.), *Radical Landscapes*, exhib. cat. Tate Liverpool. London: Tate Publishing, 2022.

Rahmann, Heike, and Jillian Walliss, *The Big Asian Book of Landscape Architecture.* Berlin: Jovis, 2020.

Raxworthy, Julian, *Overgrown: Practices between Landscape Architecture and Gardening.* Cambridge, MA: MIT Press, 2018.

Recki, Birgit, *Natur und Technik. Eine Komplikation* (*De Natura* VIII). Berlin: Matthes & Seitz Berlin, 2021.

Roth, Tobias, *Gärtnern mit Sprengstoff (1912–1929). Eine Collage.* Die Grüne Reihe, vol. 2, Berlin: SUKultur, 2022.

Sardeshpande, Mallika, et al., "Edible Urban Commons for Resilient Neighbourhoods in Light of the Pandemic", *Cities: The International Journal of Urban Policy and Planning*, vol. 109 (February 2021), https://doi.org/10.1016/j.cities.2020.103031.

Schwerin, Sophie von, et al., *Migge: The Original Landscape Designs 1910–1920 / Migge – Die originalen Gartenpläne, 1910–1920.* Basel: Birkhäuser, 2018.

Snow, Lalage, *War Gardens: A Journey through Conflict in Search of Calm.* London: Quercus Publishing, 2019.

Stauffacher Solomon, Barbara, *Green Architecture and the Agrarian Garden.* New York: Rizzoli, 1989.

Stern, Robert A. M., *Paradise Planned: The Garden Suburb and the Modern City.* New York: Monacelli Press, 2013.

Stone, Christopher D., "Should Trees Have Standing, Toward Legal Rights of Natural Objects", *Southern California Law Review*, vol. 45 (1972), pp. 450–501.

Stuart-Smith, Sue, *The Well-Gardened Mind: The Restorative Power of Nature.* New York: Scribner, 2020.

Teyssot, Georges (ed.), *The American Lawn.* Princeton, NJ: Princeton Architectural Press, 1999.

Tschumi, Christian, *Mirei Shigemori: A Rebel in the Garden.* Basel: Birkhäuser, 2007.

Tsing, Anna Lowenhaupt, *The Mushroom at the End of the World: On the Possibility of Life in Capitalist Ruins.* Princeton, NJ: Princeton University Press, 2015.

Tunnard, Christopher, *Gardens in the Modern Landscape.* Philadelphia, PA: University of Pennsylvania Press, 2014/1948.

Vogt, Günther, and Violeta Burckhardt, *Paradise Now: Die neuen Grenzen des Gartens* (*De Natura* VII), ed. Frank Fehrenbach. Berlin: Matthes & Seitz Berlin, Berlin 2021.

Waldheim, Charles, *Landscape as Urbanism: A General Theory.* Princeton, NJ: Princeton University Press, 2016.

Wall Kimmerer, Robin, *Braiding Sweetgrass: Indigenous Wisdom, Scientific Knowledge, and the Teachings of Plants.* Minneapolis, MN: Milkweed Editions, 2013.

Watson, Julia, *Lo—TEK: Design by Radical Indigenism.* Cologne: Taschen, 2019.

Wildschut, Henk, *Rooted.* self-published, 2019.

Wrede, Stuart, and William Howard Adams, *Denatured Visions: Landscape and Culture in the Twentieth Century.* New York: The Museum of Modern Art, 1991 [based on the symposium "Landscape and Architecture in the Twentieth Century", held at the Museum of Modern Art, New York, on 21/22 October, 1988].

Index

#
3D models 160–61, 195

A
Aalto, Aino 109
Aalto, Alvar 52, 109
Aben, Rob 33, 35
Acadja Fish Paddocks 191
Adirondack Chair 107
advertisements
    for chairs 110
    for fertilizers 100–103
    for garden tools 104–5
    for lawn care product 91
aerial views 50, 145, 187–88, 212
Aeschbach, Hans 103
Afghanistan 86
Africa 88–89, 108, 191, 213–14
Age of Empires (video game)
    150, 152–53
agriculture
    crop rotation 135–36
    in Ethiopia 174, 176, 179
    in greenhouses 200
    soil recovery 167
    as Tree of Life 140, 141
    see also self-sufficiency
AIDS 126, 129, 130
Albania 201
Alexander, Stephanie 206
Allegrain, Étienne 92
Allgemeine Plakatgesellschaft
    (APG) 84
allotments 16, 69, 106
    see also self-sufficiency
Aluminium 108
Ama 215
Amazon rainforest 120, 122, 166
America *see individual countries*
ammonia 100
Amsterdam 117, 119
Amun Temple 30–31
Anbauschlacht
    (cultivation battle) 84
animals 164, 166, 207
Anthropocene 162–63, 171
Antigua 138, 143
Antonelli, Paola 15
Aotearoa New Zealand 94, 185
aqueducts 29
Arbitex Dust 101
architecture and architects
    see individual persons and
    themes
Argentina 196
aristocracy 39, 92
Arles 186–87
armchairs 107
art and art projects 49, 144–49,
    196–99, 208, 213
    see also museums;
    individual artists
Ascona 16
Asia 88, 97
    see also individual countries
Aubel, Marjan van 200
Australia 58, 61, 63, 206
Austria 66, 70

B
Baan, Iwan 74, 187, 191
back gardens *see* private
    gardens
Badminton House 93
Bagh-e Shahzadeh 14, 28
Bagnaia di Viterbo 37
Bakema, Jaap 119
balconies 202
Bamako 108
Bangsar (Kuala Lumpur) 156–61
Banks, Joseph 141
Barozzi da Vignola, Giacomo 37

Barra de Guaratiba 123
Barragán, Luis 50–51
Bartram, John 141
Bartram, William 141
Basel 183, 185
Bauer, Franz 60–61
Baumann, Céline 18, 180–85
Behrens, Peter 114
Benin 191
Berlin 55, 72, 98, 114, 120, 204
Bern 84
Bian, Yujia 150, 152, 155
Bible 30, 138, 140
    see also paradisial gardens
Bilz, Friedrich Eduard 17
biodiversity 135–36, 163, 167,
    174, 215, 217
    see also ecosystems
Bodelva 198
Boki Diawe 214
bomb crater 85
books
    about design 15, 18
    about environmental policy
    95–96
    about gardens 142, 166
    about Indigenous knowledge
    190, 192–93
    by plant hunters 142–43
    about plants 117, 128
    about refugee camps 88–89
    see also Bible
Bordeaux 202
Borio, Emanuela 201
Börner, Konstantin 98
Bosch, Hieronymus 26–27
Bosco Verticale 201
botanical gardens 35, 58, 59,
    120, 161, 168
Brasília 75
Brazil 75, 120–25
Breuer, Marcel 108
Brisbane 206
Britain 61, 66, 82, 85, 92–93, 186
    see also individual cities
Brockwell Park 98
Bromme, Max 68
Brown, Lancelot Capability
    92–93
Budding, Edwin 93
Buitenzorg (now Bogor) 59
Burke, Thomas 25
Burle Marx, Roberto 17, 75,
    120–25
Burle Marx Institute 122
Burnett, Graham 209
Burnett, Mary Ann 61
Buro Happold 193

C
Calcutta 193
calendulas 137, 142
California 53, 65, 72, 74, 208
campaigns 82, 84, 85, 160
Canada 71, 75, 82
Canaletto 93
capitalist system 204
carpets 23, 29, 196–97
Carson, Rachel 95–96, 180
cathedrals 186, 189
CATPC (Cercle d'Art des
    Travailleurs de Plantation
    Congolaise) 213
Cerval, Julien de 41
chahar bagh 16
chairs 106–11, 195
Chatsworth House 217
Chatto, Beth 129, 131n2
children 83–84, 138, 140, 143, 206
    see also playgrounds
China 42–43, 150–55
Chinampas 216

Choucha refugee camp 88
Christy, Liz 80
churches 174–79
Church, Thomas 53
Cirella, Giuseppe T. 203
Clark, William 141
classification (of plants) 168–73
Clément, Gilles 18, 48, 164–67
Clifford, George 171
climate change 173, 190, 192, 196
    see also future scenarios
Clivio, Franco 104
cloth flowers 160
coasts 215
coconut 164
Cold War 72, 94
Coleridge, Samuel Taylor 141
Collins, Keith 126, 131
colonialism 16, 59, 61, 141, 213, 216
    see also decolonization
coloring (the lawn) 97
Columbus, Christoph 140, 141
commemorative gardens 71
communes 16, 208
community gardens 21, 55, 78,
    79–81, 156–61, 204–5
    see also self-sufficiency
compost 135, 156
computer games 150, 152–53
Congo 213
connecting systems (water hose)
    104
conservation walls 179
constituional rights (for plants or
    areas of nature) 185
convent gardens *see* monas-
    teries
Cook, James 141
Copacabana 75, 125
Coray, Hans 108
Cornwall 198
Cortés, Hernán 141, 142
cosmetic products 132–37
Costa, Lúcio 120, 123
Council on the Environment of
    New York City (CENYC) 81
crabgrass 95
Cranach the Elder, Lucas 24
Creation 138, 140
de' Crescenzi, Pietro 38
crop rotation 135–36
Crystal Palace 62, 70
cubism 49
Cultivation battle
    (Anbauschlacht) 84
Cupid (Cupido) 25
cycles of nature 19, 132–37, 149,
    183, 189, 192

D
Dahlem 120
dahlias 141, 142
Darmstadt 70
Darul Aman Palace 86
De 8 en Opbouw (magazine) 117
Debre Ensesa 175
decolonization 72, 75, 192, 213
Dedemsvaart 114–16, 117, 118, 119
defortification 66
Delagrive, Jean 39
Democratic Republic of the
    Congo 213
Derbyshire 195, 217
Descartes, René 171
deserts 28, 88, 214
design 15, 18
Deutsche Ammoniak-Verkaufs-
    vereinigung 100
Diallo, Cheick 108
divers 215
Dodds, Kieran 175–78
Dominica 138, 143

Donnell Garden 53
Dresden 67, 68, 72
Dr. R. Maag AG 102
Duisburg 74
Dungeness 126–31
dusting agents 101

E
Earth Day (1970) 81
East Germany *see* GDR
ecoLogicStudio 18, 19
ecosystems 63, 135–36, 164,
    166–67, 173, 180
    see also biodiversity
Ecuador 185
Eden *see* Garden of Eden
Edible Estates 205
Edible Neighbourhoods 207
Edinburgh 205
education 81, 138, 143, 160–61,
    176, 206
Egypt 30–31
Ehret, Georg Dionysius 170
Eindhoven 201
Elston, Emma 200
Emscher Park 75
Engels, Friedrich 68
England *see* Britain
Eno, Brian 20
Enright, Maginel Wright 83
environmental destruction 95,
    174, 176, 180
environmental protection 120,
    196, 210
Erfurt 72
Essen 68, 72
d'Este, Ippolito 36
Eternit Chair 111
Eternit (manufacturer) 111
Ethiopia 174–79
Europe *see individual cities and*
    countries
exhibitions 69–70, 71, 72, 75, 108
    see also museums;
    World's Fairs
expeditions (colonial) 141
explosives 101
Exposition (Switzerland) 108

F
Fach, Alexander 66
Falda, Giovanni Battista 40
farming *see* agriculture
Fauteuil 300 110
Federal Ministry of Education
    and Research 204
ferns 60–61, 135
fertilizers 100–103, 136
fibre cement 111
Fiji 63
Finland 52, 109
Firenze 168
fish ponds 36, 191, 193
Flagg, James Montgomery 82
flies 63
floating gardens 216
Flos Pavonis 18
flowers
    calendulas 137
    cloth made 160
    roses 70, 88–89, 107
    rows of 136
    wild 79, 81, 196, 217
    see also perennials
Foerster, Karl 72, 75n1, 114
Fontanet, Noël 84
food *see* agriculture
Ford, Ed 79
forests
    Bosco Verticale 201
    in Ethiopia 174–79
    for food 19, 207, 209, 213–14

Milpa Cycle 192
  rainforest 120, 122, 166
  Tokachi Millennium Forest
    210–11
  Wenlock Edge 209
Foucault, Michel 26
fountains 36
France 23, 92, 106, 108, 202, 212
  see also individual cities
Frankfurt am Main 68–69
fuel 96
Full Grown (organization) 195
Fung, Stanislaus 42
furniture 61, 106–11, 195
future scenarios
  by Bilz 17
  by Holmgren 182–83
  ideas 112–13, 173, 185, 194–217
  utopian 72, 75

G
Gailhoustet, Renée 73, 75
Games, Abram 85
Ganvie 191
Gardena (company) 104
garden books 142, 166
garden cities 67–69, 106
The Garden (film) 130
garden furniture 106–11
Garden of Eden 22–26, 28–33,
  140–41
The Garden (poem) 173n8
garden revolution 66
gardens 56–57, 112–13, 162–63,
  202–4, 216
  see also Botanical Gardens;
  community gardens;
  orchards; paradisial gardens;
  private gardens; vegetable
  and herb gardens; individual
  persons and countries
garden shows 69–70, 72
garden tools 104–5
garlic 206
Gartenschönheit (magazine) 114
Gatti, Laura 201
Gautherot, Marcel 125
GDR (German Democratic
  Republic) 55, 71–72, 75, 97, 101
Geigy (company) 102
Geng, Veronica 142
geometric structures 29, 41,
  70–71, 119
Georg III (king) 141
Gerard, John 128
Germany 72, 74, 75, 97, 106
  see also GDR; individual cities
Gerster, Georg 28
Giesecke, Annette 15, 31
Ginsberg, Alexandra Daisy 18,
  172, 173, 198–99
Girot, Christophe 30
glass 58, 61, 62–63, 202–3
Gleichenia microphylla 60–61
Gloucestershire 93
gods 46–47
Golden Age 24
gorse 129
grass (Zoysia) 96
  see also lawns; meadows
Great Britain see Britain
Great Exhibitions see World's
  Fairs
Great Green Wall 214
Greece 168, 196, 197
Green Guerillas 76–81
greenhouses 200
green space 158, 203
  see also gardens; lawns
Green Wall 214
Grossman, Sid 77
Guatemala City 19

Guerilla Gardeners 76–81
Guévrékian, Gabriel 48–49
Guhl, Willy 108, 111
Gullichsen, Harry 52, 109
Gullichsen, Maire 52, 109
Guogu, Zheng 150–55

H
Haarlem 171
Haeg, Fritz 208
Haenyeo 215
Halprin, Lawrence 65, 72
Hamelin, Pied Piper of 83
Hammerbacher, Herta 72
Haraway, Donna 173
Harlem 20, 77
Harrisburg 107
Hart, Robert 209
Hart's tongue fern 135
Hayashi, Mitsushige 210
Hearn, Maxwell K. 43
Hedblom, Marcus 98
hell 27
Hellerau (near Dresden)
  67, 68
Henselmann, Hermann 75
herbariums 168, 173
herbicides 95, 180
herbs see vegetable and herb
  gardens
Hertfordshire 21
hierarchisation (of plants)
  168–69, 173
Higashi-Hiroshima 47
High Line 146, 149
high-rise buildings 54, 201
high-voltage electricity line
  160–61
Highwood Hill 209
Hirschfeld, Christian Cay Lorenz
  66, 71
His, Andreas 102
Hitchmough, James 15, 98, 217
HIV 126, 129, 130
Hokkaido 210–11
Holland see Netherlands
Holmgren, David 180, 182–83
Hooker, William Jackson
  60–61
Horst, Arend Jan van der 119
horticulture 140, 141
hortus conclusus 17, 23, 35
House Beautiful (magazine) 53
housing projects 67–68, 73,
  202, 205
Howard, Ebenezer 68
Hugo, Victor 186
Hummelo 144, 148
humus 135
hunters (for plants) 142–43
Hyde Park 70
Hyères 48–49

I
IBA Emscher Park 75
Ignatieva, Maria 98–99
Ijsselmeer 117
Illinois 93–94
India 63, 168, 190, 193
Indigenous communities 59,
  185, 193
Indigenous knowledge 190,
  192–93, 208, 214
Indonesia 59
industrialization 67, 106, 158
insecticides 100
insects 198–99
Institute for Ecological Economy
  Research 204
Iran 14, 28–29
iron furniture 106
islands 196, 197, 215

Italy 35–37, 40, 168, 201
Ivry-sur-Seine 73, 75

J
Jacobs, Naomi 15, 31
Japan 61, 96, 203, 210–11, 215
Japanese gardens 23, 46–47
Jardin des Plantes 59
Jardines del Pedregal 50–51
Jarman, Derek 126–31
Java 59
Jefferson, Thomas 93, 141, 142
Jejú Island 215
Jekyll, Gertrude 129
journals and magazines
  De 8 en Opbouw 117
  Gartenschönheit 114
  House Beautiful 53
  New Yorker 142
  Onze Eigen Tuin 17, 117
  Species Plantarum 171

K
Kabir, Mohammad 86
Kabul 86
Kangxi (emperor) 43
Kassler, Elizabeth B. 37
Kebun-Kebun Bangsar 156–61
Kehayoglou, Alexandra 196–97
Kent 44–45, 126–31
Kew 58, 59
King James Bible 138, 140
Kircher, Athanasius 32
kitchen gardens 39, 206
Kloet, Jacqueline van der 119
knotweed 135
knowledge see Indigenous
  knowledge
Komrij, Gerrit 41
Krubsacius, Friedrich August 40
Kruse, Helmut 75n2
Kuala Lumpur 156–61
Kyoto 46–47

L
Lacaton, Anne 202
Laeuger, Max 70
Lahore 63
lakes 121, 176–77, 191, 216
land art 72
Landi Chair 108
landscape architecture and
  architects see individual
  persons and themes
language of forms 68, 70
La Quintinie, Jean-Baptiste de
  39
Latin names 142, 171
Latour, Bruno 17
lava fields 50
La Vallée 165–67
lawnmower 93, 96, 104–5
lawns 90–99, 217
learning see education
Lebanon 89
Le Corbusier 123
Lee, Thomas 107
Le Guin, Ursula K. 173
Leipzig 72
Lens 212
Levitt, Abraham 94–95
Levitt, Alfred 94–95
Levittown 94–95
Levitt, William 94–95
Lewis, Jim 15
Lewis, Meriwether 141
Liao Garden 150–52, 154–55
Liebig, Justus von 100
Ligorio, Pirro 36
lilies 141
Lindan (dusting agent) 101
Lindsay, John 78

Linnaeus, Carolus (Carl) 141, 142,
  143, 168–71
Lionni, Leo 171, 173
Lizvinova, Lubov Ninolaeuna 87
Loddiges, George 59
Loggins, Donald 80
London 62–63, 70, 85, 98, 184, 186
Lonza fertilizers 103
Lorde, Audre 183
Lo—TEK 19, 190, 192–93
Louis XIV (king) 39, 92
loungers 109
Louvre-Lens Museum 212
lower class 78, 160
  see also social housing
Lowman, Margaret 176
Luma Arles 186–87
Lusanga 213
Luxor 30–31

M
Maag AG 102
Madrid 181
magazines see journals and
  magazines
magnolias 61
Mahan 14, 28
mahogany 61
Malaysia 156–61
Mali 108
Mallet-Stevens, Robert 49
Manhattan 20, 80
Mannheim 70
manure 136
maps
  Food Forest 207
  monasteries 34–35
  New York 78
  Versailles 39
  Villa Medici 40
marigolds 137, 142
Marqueyssac 41
Marvell, Andrew 173n8
Massonnet, Henry 110
Matisse, Henri 125
matriarchal communities 215
Mattern, Hermann 72
Mayan Forest gardens 19, 192
May, Ernst 68
McHarg, Ian L. 17–18
meadows 98–99, 210–11, 217
medicinal plants 132–37
medieval 23, 35
medieval walls 66
Mehrabi, Zia 200
Melbourne 206
memorial gardens 71
Mendocino 208
Merian, Maria Sibylla 16, 18
Merritt, Andrew 182
mesh (ecological) 173
Mesopotamia 30
Metzendorf, Georg 68
MEWA (manufacturer) 108
Mexico 50–51, 216
microclimates 135, 186, 189
Middle Ages 23, 35, 168
middle classes 50, 61, 66, 78,
  93, 94
Middle East 28–31
Migge, Leberecht 68–69
Milan 201
Milos 196, 197
Milpa Cycle 19, 192
mineral fertilization 100
mining districts 75, 196–97, 212
Ministry of Education and Health
  (Brazil) 123, 125
Miró, Joan 125
Miyawaki method 189
modernism 44, 49, 70–72, 111,
  114, 117

Moerheim 114–15, 116
monasteries 17, 34–35, 168, 177
Monobloc 110
Monte Verità 16
Montreal 71, 75
Morton, Timothy 173
Mosbach, Catherine 19, 212
Moses, Robert 78
mothers 138, 140, 143
moths 63
Mowbot 105
Munanga, Charles 213
Munro, Alice 195
Munro, Gavin 195
museums
    CATPC 213
    Louvre-Lens Museum 212
    Museum of the Wind 154–55
    Museum of Wonder 152
    see also exhibitions; Vitra
    Campus

N
Nagele village 117, 119
naming of plants 142, 143, 171
National War Garden Commis-
    sion (NWGC) 82
Natur Park Südgelände 98
neighbourhoods see community
    gardens
Netherlands 114–19, 144, 148,
    200, 201, 207
network analysis 19
Neubrandenburg 72
New York City
    gardens 19, 20, 140
    Green Guerillas 76–81
    green space per individual
    203
    High Line 146, 149
    Levittown 94–95
    Victory Gardens 83
New Yorker 142
New Zealand 94, 185
Nicolson, Harold 44–45
Niemeyer, Oscar 120, 123
Nishizawa, Ryue 203
nitrogen 100
Noormarkku 52, 109
Nord-Pas-de-Calais 212
Notre-Dame 186, 189
NWGC (National War Garden
    Commission) 82
nylon 108

O
oases 28–29
oils (made of plants) 135
Olmsted, Frederick Law 93–94
Olonetzky, Nadine 24
O. M. Scott & Sons 91, 96
Ono, Haruyoshi 122
Ono, Isabela 122–23, 125
Onze Eigen Tuin (journal) 17, 117
orchards 35, 152, 195
ordering systems 168–73
Orient 23, 28–31
Ostendorf, Yasmine 200
Oudolf, Anja 144
Oudolf, Piet 119, 144–49

P
Pachamama 185
Pack, Charles Lathrop 82
painting (the lawn) 97
Palazzo Madama 35
Pandanus 122
paradisial gardens 16, 22–26,
    28–33, 86, 128, 140–41, 174
Parallel Plants 171, 173
Paris 59, 73, 75, 186, 189
parks see individual cities

Parliament of Plants 18, 181, 184
Parnis, Mollie 79
Pasteur, Louis 138, 143
Paxton, Joseph 62, 70
Pearson, Dan 210–11
pebbles 129–30
Pennsylvania 107
peppers 142
perennials 21, 114, 116, 117, 142,
    210–11
permaculture 180, 182–83
Persian gardens 14, 28–29
Peru 190
pest control 100, 101, 102
pesticides 95, 96, 102, 136, 180
Pied Piper 83
Pisa 168
Planetary Garden 18, 164–67
plantations 29, 59, 213
plants
    books about 117, 128
    classification of 168–73
    hunters for 142–43
    medicinal 132–37
    naming of 142, 171
    Parliament of Plants 18, 181,
    184
    rights for 185
    travelling ways 164
    see also flowers; trees;
    individual plants
plastic 110
playgrounds 77, 79, 97
Plumwood, Val 173
poems 131, 173n8
Pohlepp, Sascha 172
Pollan, Michael 96, 98, 173n7
Pollinator Pathmaker 198–99
polypropylene 110
Pomo 208
pools 51–53
Poppe, Nina 215
posters 82–85, 102
post-industrial parks 74, 212
Potsdam 97
potted gardens 203
Power Plant design 200
priests 174, 176, 179
private gardens 68, 70–71, 117,
    139, 148
    see also self-sufficiency
Prospect Cottage 126–31
Pteridomania 61

Q
Qanate 28

R
racism 94
Raffler, Dieter 104
railway viaduct 146, 149
rainforest 120, 122, 166
rainwater 183, 204
rationalism 171, 173
Rawsthorn, Alice 15
refugee camps 88–89
Reidy, Affonso 120
Reinagle, Philipp 25
Renaissance 24
Renaudie, Jean 73
rice farming 166
Richardson Drew, Annie Victoria
    143
Richardson, Tim 44
Riemerschmid, Richard 67, 68
Rietveld, Gerrit 116, 119
rights (for plants or areas of
    nature) 185
Rijnvliet Edible Neighbourhood
    207
Rio de Janeiro 75, 121–25
rivers 185, 196–97

Riverside 93–94
robotic 105
rocking chairs 108, 111
Rome 40
rooftop gardens 123, 125, 183, 200
root fly 101
Rosellini, Ippolito 30–31
roses 70, 88–89, 107
Roth, Tobias 25
Rousseau, Jean-Jacques 66, 171
Royal Botanic Gardens 58, 59
rubber 59
rugs 23, 29, 196–97
Ruhr region 74, 75
Russo, Alessio 203
Ruys, Mien 17, 75n1, 114–19

S
Sackville-West, Vita 44–45
Safdie, Moshe 71, 75
Sahel region 214
Salas Portugal, Armando 50
Salmon Creek Farm 208
Salomonson, Hein 119
San Francisco 72, 74
sanitary conditions 66, 67
Santa Cruz River 196–97
São Paulo 75, 125
Sato-umi 215
School Garden Army 83
schools 83, 183, 206
Schreber gardens 16, 69, 106
    see also self-sufficiency
Schreber, Gottlieb Moritz 16
Schwäbisch Gmünd 132–37
science fiction 173
Scotland 205
Scott & Sons 91, 96
Scott, Frank J. 94
sculptures 127, 130–31, 213
seakale 128, 129
Sea Ranch 65, 72, 74
seed bombs 76, 81
seeds 164
Ségou (rocking chair) 108
Sek San, Ng 19, 156–61
self-sufficiency
    Edible Estates 205
    food forests 19, 207, 209,
    213–14
    gardens for 20, 54, 68–69, 82,
    84, 156–61
    kitchen gardens 39, 206
    see also vegetable and herb
    gardens
Senegal 214
Sennefer 30–31
sexual classification (of plants)
    170, 171
Shabib, Rashid bin 29
Shepherd, Harry 85
Shigemori, Mirei 46–47
Shintani, Midori 210–11
Shropshire 209
Silva Tarouca, Ernst Graf 117
Simonovka 87
Singapore 94
Sissinghurst Castle 44–45
Sítio Roberto Burle Marx 121, 123
Sítio Santo Antônio da Bica 123
Smart Oath of Understanding
    (SOOU) 193
Smet, Bas 19, 186–89
Smith Barton, Benjamin 142
Smythe, Frank 142–43
Snow, Lalage 86–87
social housing 67–68, 73, 202,
    205
soils 135–36, 179
Solander, Daniel Charles 141
Sooley, Howard 127–31
South Korea 215

space and spatial structure
    public 74–75, 156, 159–62
    rooftops 183, 200
    symbolic 41, 92
    vertical gardens and forests
    201, 204
    winter gardens 202
    see also future scenarios;
    urban developments
spa gardens 106
Spain 181
Species Plantarum (magazine)
    171
sponge cities 183, 189
Spray Rose 89
Sprenger, Astrid 133
STAMP (company) 110
Stauffacher Solomon, Barbara
    16, 39
steel and steelworks 74, 108
Steyerl, Hito 173
St. Gall 34
Stone, Christopher 185
stones 46–47, 129
street map (New York) 78
structuralism 119
Stuart-Smith, Sue 20, 21
Stuart-Smith, Tom 21
Studio Burle Marx & Cia. Ltda. 122
Stuttgart 72
suburban communities 71, 78,
    93, 94
Suriname 16, 18
sustainability 135–36, 192–93, 210
Sweden 99
sweet potatoes 142
swimming pools 51–53
Swinton, Tilda 126
Switzerland 16, 34, 82, 84, 108
Systema Naturae 169

T
Tabacow, José 122
Tana Lake 176–77
Tang Yifen 42
tapestries 196–97
Tasmania 206
Taut, Bruno 114
Taut, Max 114
taxonomy 168–73
technologies 190, 192–93
Temple of Amun 30–31
Terbol 89
terraced houses 67
Tewahedo Church 174, 176
Thoma, Hans 31
Thyssen steelworks 74
Tirana 201
Tivoli 36
Tokachi Millennium Forest 210
Tokyo 203
Tolou Keur garden 214
tomatoes 168
tombs 30
tools for gardening 104–5
traditions
    biblical 31, 140
    of the commons 78
    functional 48
    japanese 203, 210–11
    Lo—TEK 190, 192–93
    Malian 108
transports 58–59, 61, 63
travelling (animals and plants) 164
trees 140, 141, 185, 195, 201
    see also forests
Treister, Suzanne 173
Tsing, Anna Lowenhaupt 183
Tuinen Mien Ruys 114–16, 117,
    118, 119
Tunisia 88
Tunnard, Christopher 53

Turin 35
twinleaf 142

U
Ukraine 87
UNESCO World Heritage sites 123
United Kingdom *see* Britain
United States of America *see* USA
University of Malaya 161
upper class 61, 78
Uppsala 99
urban developments 66–69,
    70–81, 117–19, 125, 185
urea-based fertilizer 101
USA (United States of America)
    garden furniture 107
    Guerilla Gardeners 76–81
    land art 72
    lawns 93, 94–96
    School Garden Army 83
    self-sufficiency 82
    Wardian cases 61
    see also individual cities
utopias 72, 75
Utrecht 207

V
Valk, Gerard 17, 18
Vassal, Jean-Philippe 202
VEB Berlin-Chemie 101
vegetable and herb gardens 23,
    34–35, 39, 204
    see also self-sufficiency
Vermont 139, 141–42
Versailles 39, 92
vertical gardens and forests 201,
    204
Vézac 41
Vichy 106
Victory Gardens 78, 82, 83, 85
Vienna 66, 70
Villa d'Este 36
Villa Lante 37
Villa Mairea 52, 109
Villa Medici 40
Villa Noailles 48–49
Vitra Campus 144, 145–47,
    188, 189

W
Wagner, Martin 71
walls 66, 179, 214
Wang, Hui 43
Wardian cases 58–59, 61, 63
Ward, Nathaniel Bagshaw
    58–59, 61, 63
wars *see* Cold War; World War
    I and II
Washington, George 93
wasps 63
Wassie Eshete, Alemayehu
    174–79
water
    to cool down 186
    cycles 183, 189, 204
    for lawns 96
    in Liao Garden 155
    in Persian gardens 28–29
    in Villa d'Este 36
wastewater 193
water hose connecting system 104
water lily 70
Watson, Julia 19, 190–93
weeds 92, 166
Weil am Rhein 144, 145–47,
    188, 189
Weleda (company) 132–37
Wells, H. G. 173
Wenlock Edge 209
West Gojam 176
West Indies 138, 140, 141, 143
Westport Chair 107

Whanganui River 185
White Cube (museum) 213
white fly 63
Wiesbaden 66–67
wildflowers 79, 81, 196, 217
Wildschut, Henk 88–89
Williams, Henry T. 61
Wines, James 54
winter gardens 202
Wirksworth 195
Wit, Saskia de 33, 35
Woglum, Russel 63
women communities 215
Woolmington, Rob 139
word origins 24
Wordsworth, William 141
World Health Organisation
    (WHO) 203
World Heritage sites (UNESCO)
    123
World's Fairs
    London (1851) 62–63, 70
    Montreal (1967) 75
World War I and II 78, 82–85

X
Xochimilco 216

Y
Yangjiang 150–52, 154–55
youth 83–84
    see also playgrounds

Z
Zais, Christian 66–67
Zege 178
Zimmermann, Peter 55
Zoysia grass 96
Zurich 84, 108

Image Credits

14 → Photo: S.H. Rashedi

16 → © Fondazione Monte Verità, Ascona, Fondo Harald Szeemann

17 → Public Domain

18 ↑ Niedersächsische Staats- und Universitäts- bibliothek Göttingen, GR 2 ZOOL VI, 3904 RARA

18 ↓ © Vitra Design Museum

19 → © ecoLogicStudio

20 → © Green City Force

21 → Photo: Jiaji Wu, Courtesy of Sue Stuart-Smith

24 → Nasjonalmuseet Oslo

25 → Wellcome Collection

26– → © Photographic Archive
27 Museo Nacional del Prado

28 → Photo: Georg Gerster, © Estate Georg Gerster

29 → The Metropolitan Museum of Art, The James F. Ballard Collection, Gift of James F. Ballard, 1922

30 → Universitätsbibliothek Heidelberg

31 → © Kunsthalle Karlsruhe

32 → © British Library

33 → © Sammlung Städel Museum Frankfurt

34 → © St. Gallen, Stiftsbibliothek

35 ↑ © Torino, Palazzo Madama – Museo Civico d'Arte Antica. Courtesy Fondazione Torino Musei, photo: Giorgio Perottino, 2018

35 ↓ © Torino, Palazzo Madama – Museo Civico d'Arte Antica. Courtesy Fondazione Torino Musei, photo: Giorgio Perottino, 2018

36 → Photo: Iwan Baan, 2016

37 → Bibliotheca Hertziana Rom – MPI für Kunst- geschichte, photo: Marcello Leotta

38 → Bibliothèque nationale de France, Ms-5064

39 ↑ Technische Informationsbibliothek (TIB), Sammlung Albrecht Haupt, gr F A 1:1

39 ↓ Photo: Paris Histoire

40 ↑ © Rijksmuseum, Amsterdam

40 ↓ SLUB Dresden / Deutsche Fotothek

41 → © Photo: Romain Laprade, 2020

42 → © The Trustees of the British Museum

43 → The Metropolitan Museum of Art, New York

44– → © National Trust
45 Images/Jonathan Buckley

46 → Photo: Christian Lichtenberg

47 ↑ Photo: Christian Lichtenberg

47 ↓ Photo: Christian Lichtenberg

48 → © Collection Villa Noailles

49 → © Courtesy of University of Illinois Archives, Gabriel Guevrekian Papers, image 024

50 ↑ Luis Barragán, Armando Salas Portugal (Photo) © Barragan Foundation / VG Bild-Kunst, Bonn 2023

50 ↓ Fondo Aerofotográfico Acervo Histórico Fundación ICA, A.C., Número de Control FAO_01_009273, photo: Compañia Mexicana Aerofoto

51 → LIGA – Archivos in Mexico City, photo: Roberto and Fernando Luna

52 → Alvar Aalto Foundation, photo: Gustaf Welin, 1939

53 → Photo © Charles A. Birnbaum, courtesy of Cultural Landscape Foundation

54 → Sammlung Jonathan Holtzman © 2022 James Wines

55 → Bundesarchiv, Bild 183-1982-0807-019 / photo: Peter Zimmermann

58 → © The Board of Trustees of the Royal Botanic Gardens, Kew

59 ↑ © The Board of Trustees of the Royal Botanic Gardens, Kew

59 ↓ © Collection Nationaal Museum van Wereldculturen. Coll.no. TM-10010760

60 → © Royal College of Physicians of Edinburgh

61 ↑ Album / Alamy Stock Photo

61 ↓ Illustration from Henry T. Williams, *Window Gardening: Devoted Specially to the Culture of Flowers and Ornamental Plants, for Indoor Use and Parlor Decoration,* New York, 1874

62 → Lordprice Collection / Alamy Stock Photo

63 → Photo from Russell Woglum, *Report of a Trip to India and the Orient in Search of the Natural Enemies of the Citrus White Fly,* Washington, D.C., 1913

65 → © Lawrence Halprin

66 → Hessisches Haupt- staatsarchiv, Best. 3011/1 Nr. 1715 V

67 → © bpk / Kunstbibliothek, SMB, VG Bild-Kunst, Bonn 2023

68 ↑ © Town & Country Planning Association

68 ↓ Archiv für Schweizer Landschaftsarchitektur ASLA

69 → © Institut für Stadt- geschichte Frankfurt am Main (ISG FFM), S7A/1998-22518, n.n.

70 → Image from *Architektonische Rundschau,* no. 11, 1907, p. 87; courtesy of Universitätsbibliothek Heidelberg

71 → © Safdie Architects

73 → Lorenzo Zandri © 2017 (Ivry-sur-Seine, FR)

74 ↑ Photo: © Iwan Baan

74 ↓ Jochen Tack / Alamy Stock Photo

77 → Sid Grossman (1915–1955) for Federal Art Project. Museum of the City of New York

78 → P. Maverick, *This map of the City of New York and island of Manhattan as laid out by the commissioners... by their most obedient servant Wm. Bridges,* 1811, 24 × 92 cm, NYHS Maps – M6.1.4 / inventory no. M002414.1, New-York Historical Society Library, image number 34595; © New-York Historical Society

79 → © Library of Congress, Prints and Photographs Division, Washington, D.C. [LC-USZ62-122639], photo: Ed Ford

80 ↑ © Donald Loggins

80 ↓ © Donald Loggins

81 → © Image courtesy of Green Guerillas

82 → © Library of Congress, Prints and Photographs Division, Washington, D.C. [LC-USZC4-10234]

83 ↑ © Library of Congress Prints and Photographs Division Washington, D.C. [LC-DIG- ppmsca-53320]

83 ↓ © Library of Congress Prints and Photographs Division Washington, D.C. [LC-DIG-fsa-8d35488 DLC], photo: Edward Meyer

84 ↑ © Plakatsammlung der Schule für Gestaltung Basel

84 ↓ © Keystone/ Photopress-Archiv/Str

85 ↑ © Library of Congress, Prints and Photographs Division, Washington, D.C. [LC-USE6-D-009202-1]

85 ↓ © Digital image, The Museum of Modern Art, New York/Scala, Florence

86 → © Photo: Lalage Snow

87 → © Photo: Lalage Snow

88 → © Photo: Henk Wildschut

89 → © Photo: Henk Wildschut

91 → © Vitra Design Museum, photo: Andreas Sütterlin

92 → Bridgeman Images

93 → bpk / RMN - Grand Palais / Gérard Blot

94 → © Frederick Law Olmsted Society

95 → © Photo: Levittown Public Library

96 ↑ Reprinted by permission of Houghton Mifflin Harcourt Publishing Company

96 ↓ © Vitra Design Museum, photo: Andreas Sütterlin

97 → Bundesarchiv, Bild 183-C0619-0009-001, photographer: Heinz Junge

98 → Photo: Konstantin Börner

99 → Photo: Maria Ignatieva, 2019

100 → © Vitra Design Museum, photo: Andreas Sütterlin

101 ↑ Courtesy of BASF

101 ↓ Deutsches Kleingärtnermuseum Leipzig e.V.

102 ↑ Zurich University of the Arts / Museum für Gestaltung Zürich / Poster Collection

102 ↓ Courtesy of Novartis Archive, photo: Andreas Sütterlin

103 → Plakatsammlung Schule für Gestaltung Basel

104 ↑ Courtesy of Gardena

104 ↙ Courtesy of Gardena

104 ↘ Courtesy of Deutsche Fotothek, Stiftung Deutsches Design Museum (unknown photographer)

105 → © Vitra Design Museum

106 ↑ © Vitra Design Museum, photo: Jürgen Hans

106 ↓ From Georg Himmelheber, *Möbel aus Eisen: Geschichte, Formen, Techniken.* Munich, 1996, p. 376. Courtesy of Georg Himmelheber

107 ↑ Smithsonian Institution, Archives of American Gardens, McFarland company collection 1899–1974

107 ↓ © Vitra Design Museum, photo: Jürgen Hans

108 ↑ © Vitra Design Museum, photo: Jürgen Hans

108 ↓ © Vitra Design Museum, photo: Andreas Sütterlin

109 ↑ © Vitra Design Museum, photo: Jürgen Hans

109 ↓ © Artek Collection / Alvar Aalto Museum

110 ↑ © Vitra Design Museum, photo: Jürgen Hans

110 ↓ © STAMP, photo © Vitra Design Museum

111 ↑ Courtesy of Eternit (Schweiz) AG, Niederurnen

111 ↓ © Vitra Design Museum, photo: Jürgen Hans

115 → © Stichting Tuinen Mien Ruys

116 ↑ © Buro Mien Ruys

116 ↙ © Stichting Tuinen Mien Ruys

116 ↘ © Buro Mien Ruys

117 → © Buro Mien Ruys

118 ↑ © Buro Mien Ruys, drawing: Anet Scholma

118 ↙ © Buro Mien Ruys

118 ↘ © Buro Mien Ruys

119 ↖ Photo: Leo den Dulk

119 ↗ © Stichting Tuinen Mien Ruys

121 → Photo: José Tabacow, Collection: Iphan

122 ↖ Sítio Roberto Burle Marx Collection

Acknowledgements

The editorial and curatorial team would like to express their gratitude to all artists, lenders, licensors, and advisors involved in the exhibition. Among them are:

Alexandra Kehayoglou Studio: Jose Huidobro, Alexandra Kehayoglou

Archive of the Hochschule für Gestaltung Ulm: Stefanie Dathe, Christiane Wachsmann

Barragan Foundation: Martin Josephy, Federica Zanco

BASF Archive: Isabella Blank-Elsbree

Basilisk Communications: James Mackay

Botanischer Garten und Botanisches Museum Berlin: Sylke Gottwald, Susanne Feldmann

Burle Marx Institute: Cecilia Ewbank, Tatiana Leiner, Isabela Ono

Buro Bas Smets: Jacopo Fochi, Bas Smets

Buro Mien Ruys: Anet Scholma

Céline Baumann Studio: Céline Baumann, Juan Brunetti

Computerspielemuseum Berlin: Nicole Hanisch

Concrete Jungle: Katherine Kennedy

Dan Pearson Studio: Huw Morgan, Dan Pearson

Deutsches Gartenbau-museum Erfurt: Alieda Halbersma

Deutsches Kleingärtner-museum Leipzig e.V.: Caterina Paetzelt, Alexandra Uhlisch

Dreiländermuseum, Lörrach: Arne Gentzsch, Ulrike Konrad

ecoLogicStudio: Claudia Pasquero, Marco Poletto, Alessandra Poletto

Edible Estates: Greig Robertson

Emergence Magazine: Devin Tellatin, Emmanuel Vaughn-Lee

Estate Georg Gerster AirPhotography: Anya von Schweinitz-Calonder

Eternit AG: Gabriella Gianoli, Julia Kessler

Full Grown: Alice Munro, Gavin Munro

Garden Museum London: Emma House, Philip Norman, Alice Ridgway

Green City Force: Tonya Gayle

Human Activities: Nikki Omes

Institute for Ecological Economy Research: Antonia Sladek

Institute for the History of Frankfurt: Tobias Picard

Instituto do Patrimônio Histórico e Artístico Nacional | IPHAN: Jéssica Santana, Rafael Zamorano Bezerra

Jardins de Marqueyssac: Nathalie Bapst, Olivia Jewell-Harrison

Kebun-Kebun Bangsar: Joanne Mun, David Tan

Keith Collins Will Trust: Amanda Wilkinson

Lacaton & Vassal: Anne Lacaton, Jean-Philippe Vassal

Lo—TEK: Luiza Livingston, Julia Watson

Louisiana: Christian Lund

Marjan van Aubel Studio: Eva Trum, Marjan van Aubel

Mosbach Paysagistes: Catherine Mosbach

Museum für Gestaltung Zürich: Patrizia Baldi, Renate Menzi

Mynd: Natasha Kraal

Nieuwe Instiuut: Behrang Mousavi, Emily Wijns

Novartis Archive: Walter Dettwiler, Florence Wicker

Schule für Gestaltung Basel, Plakatsammlung: Philipp Messner

Archiv für Schweizer Landschaftsarchitektur ASLA: Hansjörg Gadient, Simon Orga, Sophie von Schwerin

Something & Son: Andrew Merritt

Special Collections Wageningen: Anita Dijkstra, Anneke Groen

Stadtarchiv Weil am Rhein: Thilo Baumgartner

Stadt Weil am Rhein Landwirtschaftsmuseum: Simone Kern, Peter Spörrer

Stefano Boeri Studio: Stefano Boeri, Siriana Guoyin, Anastasia Kucherova

Stephanie Alexander Kitchen Garden Foundation: Stephanie Alexander, Nikki Anderson

Stichting Tuinen Mien Ruys: Monique Bruins Slot

Stiftung für Kunst, Kultur und Geschichte, Winterthur: Julian Cech, Jasmin Eckhardt, Severin Rüegg

Studio Daisy Ginsberg: Freire Barnes, Alexandra Daisy Ginsberg

Studio Dots: Laura Drouet, Olivier Lacrouts

Studio Rustemeyer: Thomas Rustemeyer

The Blink Fish Film Productions: Paolo Soravia

Tokachi Millennium Forest: Midori Shintani

Town & Country Planning Association: Katy Lock

Universitätsbibliothek Basel: Bettina Braun, Alice Keller

Vitamin Creative Space: Wei Zhang, Wenzhe Zhu

Wylie Agency: Tucker Smith

Leila Ashtari

Iwan Baan

Lucky Jeffrey Brown

Jan Buchczik

Graham Burnett

Saskia de Wit

Kieran Dodds

Siew Chew Yue

Franco Clivio

Laurie Cluitmans

Stefani Fricker

Volker Gehrmann

Zheng Guogu

Fritz Haeg

Dana Halprin

James Hitchmough

Bettina Jaugstetter

Jamaica Kincaid

Romain Laprade

Monika Lauria

Christian Lichtenberg

Anni Lionni

Donald Loggins

Pedro Mascaro

Ryue Nishizawa

Andreas Nutz

Nadine Olonetzky

Piet Oudolf

Thomas Piper

Nina Poppe

Dieter Raffler

Tim Richardson

Lalage Snow

Howard Sooley

Sue Stuart-Smith

José Tabacow

Suzanne Tóth-Pál

Christian Tschumi

Henk Wildschut

James Wines

Suzan Wines

Rob Woolmington

For their insightful contributions to this book, we would like to thank Céline Baumann, Yujia Bian, Gilles Clément, Lisa Dabscheck, Leo den Dulk, Kieran Dodds, Jochen Eisenbrand, Zheng Guogu, Maria Heinrich, Luke Keogh, Jamaica Kincaid, Kris Kozlowski Moore, Marten Kuijpers, Donald Loggins, Christoph Miler, Isabela Ono, Hanno Rauterberg, Ng Sek San, Bas Smets, Howard Sooley, Astrid Sprenger, Oliver Sukrow, Alemayahu Wassie Eshete, and Julia Watson.

We thank Henrike Büscher for the editorial coordination and our team of content editors, translators, copy-editors, and proofreaders led by Kirsten Thietz and Amanda Gomez and including Nina Hausmann.

Our gratitude to Lorenz Klingebiel and Dominik Krauss for the inspired book design and the exhibition graphics, to Formafantasma for the meticulous exhibition design, and of course to all present and former colleagues at the Vitra Design Museum and the Nieuwe Instiuut who collaborated on this exhibition project with dedication, expertise, and enthusiasm.

Last but not least, we would like to thank our families and friends for actively supporting us with their wisdom and many warm meals.

Colophon

This book is published on the occasion of the exhibition
Garden Futures: Designing with Nature

Vitra Design Museum, Weil am Rhein
25 March 2023 – 3 October 2023

Design Museum Helsinki & Museum of Finnish
Architecture, Helsinki
10 November 2023 – 1 April 2024

Vandalorum, Värnamo
27 April 2024 – 13 October 2024

Nieuwe Instituut, Rotterdam
November 2024 – March 2025

V&A Dundee
April 2025 – December 2025

Further venues are planned.

Publishers: Vitra Design Museum, Wüstenrot Foundation
Concept: Viviane Stappmanns, Mateo Kries
Editors: Viviane Stappmanns, Nina Steinmüller,
Carolina Maddè
Editorial Management: Henrike Büscher, Kirsten Thietz
Copy-editing and Proofreading: Amanda Gomez,
Laura Preston, Nina Hausmann (English),
Kirsten Thietz, Nina Hausmann (German)
Translations: Herwig Engelmann, Claudia Kotte,
Martin Hager (German), Bronwen Saunders,
Lisa Schons (English)
Index: Jutta Mühlenberg
Image Rights: Wen Bi, Carolina Maddè, Nina Steinmüller

Design: Lorenz Klingebiel and Dominik Krauss
Project Management: Esther Schröter
Production: Judith Brugger
Distribution: Pinar Yildiz
Pre-press: GZD Media GmbH, Renningen
Printing: DZA Druckerei zu Altenburg GmbH, Altenburg
Cover fabric: Peyer Duchesse Offset white, 170g
Paper: LonaOffset 140 gsm (FSC and EU Ecolabel certified)
Typefaces: Modern Gothic by Malte Bentzen, AllCaps;
ABC Marist by Seb McLauchlan, Dinamo

First published by the Vitra Design Museum and the
Wüstenrot Foundation

Vitra Design Museum
Charles-Eames-Straße 2
79576 Weil am Rhein
Germany
verlag@design-museum.de

Wüstenrot Foundation
Gemeinschaft der Freunde Deutscher Eigenheimverein e.V.
Hohenzollernstraße 45
71638 Ludwigsburg
Germany
info@wuestenrot-stiftung.de

Printed and bound in Germany
© Vitra Design Museum 2023

ISBN (English edition): 978-3-945852-53-8
ISBN (German edition): 978-3-945852-52-1

Exhibition

Curators: Viviane Stappmanns, Marten Kuijpers
Co-Curator: Nina Steinmüller
Curatorial Assistance: Wen Bi, Maria Heinrich,
Wietske Nutma
Initial Research: Erika Pinner
Project Management: Carolina Maddè
Design: Formafantasma
Graphics: Lorenz Klingebiel and Dominik Krauss
Image and Film Rights: Wen Bi, Carolina Maddè,
Nina Steinmüller
Technical Director: Stefani Fricker
Exhibition Development: René Herzogenrath, Erika Müller
Senior Art Technicians: Olaf Krüger, Niels Tofahrn,
Martin Wittwer, Daniel Dressel, Bernd Nickel
Conservation: Susanne Graner, Lena Hönig
Press and Public Relations, Marketing: Johanna Hunder,
Jan-Marcel Müller, Maximilian Kloiber
Partnerships: Jasmin Zikry
Education and Public Programme: Katrin Hager, Tom Nieke,
Coline Ormond
Exhibition Tour: Cora Harris, Dominique Jahn
Publications: Esther Schröter, Judith Brugger, Pinar Yildiz
Registrar: Sarah Aubele
Archive: Andreas Nutz
Visitor Experience: Rebekka Nolte
Visitor Services: Felix Ebner, Katharina Herrmann
Museum Shops: Florian Otterbach

Vitra Design Museum
Director: Mateo Kries
COO / Deputy Director: Sabrina Handler
Head of Finance: Heiko Hoffmann

Nieuwe Instituut
General and Artistic Director: Aric Chen
Business Director: Josien Paulides
Head of Programme: Flora van Gaalen

Wüstenrot Foundation
Managing Director: Prof. Philip Kurz
Head of Arts & Culture: Verena Krubasik
Project Manager: Laura Sophie Puin

An exhibition by the Vitra Design Museum, the Wüstenrot
Foundation, and the Nieuwe Instituut

Global Partner   Partner

Thanks to